中国
城乡建设统计年鉴

中华人民共和国住房和城乡建设部 编

Ministry of Housing and Urban-Rural Development, P.R.CHINA

2020

China
Urban-Rural
Construction
Statistical
Yearbook

U0351828

中国统计出版社
China Statistics Press

图书在版编目（CIP）数据

中国城乡建设统计年鉴. 2020 = China Urban-Rural Construction Statistical Yearbook 2020：汉英对照 / 中华人民共和国住房和城乡建设部编. -- 北京：中国统计出版社, 2021.9

ISBN 978-7-5037-9627-2

Ⅰ.①中… Ⅱ.①中… Ⅲ.①城乡建设－统计资料－中国－2020－年鉴－汉、英 Ⅳ.①TU984.2-54

中国版本图书馆 CIP 数据核字(2021)第 172693 号

中国城乡建设统计年鉴—2020

作　　者/ 中华人民共和国住房和城乡建设部
责任编辑/ 钟　钰
封面设计/ 张　冰
出版发行/ 中国统计出版社有限公司
地　　址/ 北京市丰台区西三环南路甲 6 号　邮政编码/100073
电　　话/ 邮购（010）63376909　书店（010）68783171
网　　址/ http://www.zgtjcbs.com
印　　刷/ 河北鑫兆源印刷有限公司
经　　销/ 新华书店
开　　本/ 890mm×1240mm　1/16
字　　数/ 432 千字
印　　张/ 14.5
版　　别/ 2021 年 9 月第 1 版
版　　次/ 2021 年 9 月第 1 次印刷
定　　价/ 168.00 元

如有印装错误，本社发行部负责调换。

《中国城乡建设统计年鉴—2020》

编委会和编辑工作人员

一、编委会

主　　任：姜万荣

副 主 任：胡子健

编　　委：（以地区排名为序）

虞京平　杨　琦　吴秉军　王　伟　桑卫京　吴　铁　符里刚　翟顺河
刘建华　韩有明　李海洋　华天舒　邢文忠　柴　冠　赖晓峰　刘晓东
姚　凯　陈浩东　刘大威　应柏平　张　奕　刘孝华　高　宇　江新洪
李道鹏　王玉志　李新怀　张秀梅　张　弘　谈华初　易小林　彭国安
刘耿辉　叶　云　汪夏明　宋　祎　陈光宇　冯　赵　陈福均　杨　搏
陈　勇　曹鸣凤　柳明林　赵志勇　格　桑　姜月霞　茹广生　杨福波
李兰宏　白宗科　马汉文　黎向群　蔡启明

二、编辑工作人员

总 编 辑：胡子健（兼）

副总编辑：吴文君　辛芸娜

编辑人员：（以地区排名为序）

张　文　樊　荣　诸葛绪光　石　林　胡　斌　张志嵩　张利兵　付兴华
刘占峰　张　伟　李芳芳　封　刚　李真超　闫玉变　杨　婧　张海星
邵丽峰　梁立方　边晓红　杨宝峰　张亭亭　沈晓红　郭　赟　王光辉
林　岩　刘　金　郑斯琪　刘海臣　于在洋　王志成　周　宇　孙莉利
谷洪义　杜金芝　韩　宇　苑晓东　赵　哲　张钧誌　鲁　超　路文龙
王　青　张　力　王佳剑　陈明伟　王　畅　王山红　柏一韬　何青峰
邱亦东　余成海　沈昆卉　余　燕　陈文杰　王少彬　王晓霞　黄新华
林　伟　涂　莉　冯兰成　姚　坡　麻鹏飞　刘二鹏　郭彩文　吴学英
韩文超　曹　键　张　雷　陈建军　凌小刚　禹滋柏　周　旭　蒋　琳
杨爱春　梁　钦　尹清源　赵鹏凯　周倩仪　孙　玮　陆　翔　吴　茵
刘　哲　黄玉清　高　磊　罗　翼　叶　茂　廖　楠　熊　风　叶晓璇
刘宇飞　林　琳　董小星　鲁长亮　彭　里　安志勤　李平辉　王新宇
白庆武　和　兴　付　聪　熊艳玲　罗桑次仁　德庆卓嘎　边　珍　胡亚红
杨　莹　王光辉　毕东涛　刘　鹏　杨敏政　刘　波　冶姿祎　朱燕敏
李　崑　李　军　叶维晶　张少艾　杨　帆　王舒馨　张静宇　李璠璠
李雪娇

China Urban-Rural Construction Statistical Yearbook—2020

Editorial Board and Editorial Staff

编辑说明

一、为全面反映我国城乡市政公用设施建设与发展状况，方便国内外各界了解中国城乡建设全貌，我们编辑了《中国城乡建设统计年鉴》《中国城市建设统计年鉴》和《中国县城建设统计年鉴》中英文对照本，每年公布一次，供社会广大读者作为资料性书籍使用。

二、本年鉴的统计范围

设市城市的城区：市本级（1）街道办事处所辖地域；（2）城市公共设施、居住设施和市政公用设施等连接到的其他镇（乡）地域；（3）常住人口在3000人以上独立的工矿区、开发区、科研单位、大专院校等特殊区域。

县城：（1）县政府驻地的镇、乡或街道办事处地域（城关镇）；（2）县城公共设施、居住设施和市政公用设施等连接到的其他镇（乡）地域；（3）常住人口在3000人以上独立的工矿区、开发区、科研单位、大专院校等特殊区域。

村镇：政府驻地的公共设施和居住设施没有和城区（县城）连接的建制镇、乡和镇乡级特殊区域。

三、2020年底全国境内31个省、自治区、直辖市，共有685个设市城市，1483个县（含自治县、旗、自治旗、林区、特区。下同），21157个建制镇，8809个乡（含民族乡、苏木、民族苏木。下同），50.9万个行政村。

四、本年鉴根据各省、自治区和直辖市建设行政主管部门上报的2020年城乡建设统计数据编辑，由城市（城区）、县城、村镇三部分组成：

（一）城市（城区）部分，统计了684个城市、3个特殊区域。3个特殊区域包括吉林省长白山保护开发区、陕西省杨凌区和宁夏回族自治区宁东。

（二）县城部分，统计了1479个县和16个特殊区域。河北省邢台县、沧县，山西省泽州县，辽宁省抚顺县、铁岭县，河南省安阳县，新疆维吾尔自治区乌鲁木齐县、和田县等8个县，因与所在城市市县同城，县城部分不含上述县城数据，数据含在其所在城市中；福建省金门县暂无数据资料。

16个特殊区域包括河北省曹妃甸区、白沟新城，山西省云州区、太谷区，黑龙江省加格达奇区，湖北省神农架林区，湖南省望城区、南岳区、大通湖区，海南省洋浦开发区，贵州省六枝特区、水城区，云南省昆明阳宗海风景名胜区，宁夏回族自治区红寺堡区，青海省海北州府西海镇、大柴旦行委。

（三）村镇部分，统计了 18822 个建制镇、8876 个乡、447 个镇乡级特殊区域和 236.3 万个自然村（其中村民委员会所在地 49.3 万个）。

五、本年鉴数据不包括香港特别行政区、澳门特别行政区以及台湾省。

六、本年鉴中除人均住宅建筑面积、人均日生活用水外，所有人均指标、普及率指标均以户籍人口与暂住人口合计为分母计算。

七、本年鉴中"空格"表示该项统计指标数据不足本表最小单位数、数据不详或无该项数据。

八、本年鉴中部分数据合计数或相对数由于单位取舍不同而产生的计算误差，均没有进行机械调整。

九、为促进中国建设行业统计信息工作的发展，欢迎广大读者提出改进意见。

EDITOR'S NOTES

1. *China Urban-Rural Construction Statistical Yearbook*, *China Urban Construction Statistical Yearbook* and *China County Seat Construction Statistical Yearbook* are published annually in both Chinese and English languages to provide comprehensive information on urban and rural service facilities development in China. Being the source of facts, the yearbooks help to facilitate the understanding of people from all walks of life at home and abroad on China's urban and rural development.

2. Coverage of the statistics

Urban Areas: (1) areas under the jurisdiction of neighborhood administration; (2) other towns (townships) connected to urban public facilities, residential facilities and municipal utilities; (3) special areas like independent industrial and mining districts, development zones, research institutes, and universities and colleges with permanent residents of 3000 and above.

County Seat Areas: (1) towns and townships where county governments are situated and areas under the jurisdiction of neighborhood administration; (2) other towns (townships) connected to county seat public facilities, residential facilities and municipal utilities; (3) special areas like independent industrial and mining districts, development zones, research institutes, and universities and colleges with permanent residents of 3000 and above.

Villages and Small Towns Areas: towns, townships and special district at township level of which public facilities and residential facilities are not connected to those of cities (county seats).

3. There were a total of 685 cities, 1483 counties (including autonomous counties, banners, autonomous banners, forest districts, and special districts), 21157 towns, 8809 townships (including minority townships, Sumus and minority Sumus), and 509 thousand administrative villages in all the 31 provinces, autonomous regions and municipalities across China by the end of 2020.

4. The yearbook is compiled based on statistical data on urban and rural construction in year 2020 that were reported by construction authorities of provinces, autonomous regions and municipalities directly under the central government. The yearbook is composed of statistics for three parts, namely statistics for cities (urban districts), county seats, and villages and small towns.

(1) In the part of Cities (Urban Districts), data are from from 684 cities, 3 special zones. Changbai Mountain Protection and Development District in Jilin Province, Yangling District in Shaanxi Province and Ningdong in ningxia Autonomous Region are classified as city.

(2) In the part of County Seats, data is from 1479 counties, 16 special zones and districts. Data from 8 counties including Xingtai and Cangxian County in Hebei Province, Zezhou County in Shanxi Province, Fushun and Tieling County in Liaoning Province, Anyang County in Henan Province, and Urumqi and Hetian County in Xinjiang Uygur Autonomous Region are included in the statistics of the respective cities administering the above counties due to the identity of the location between the county seats and the cities; data on Jinmen County in Fujian Province has not been available at the moment.

16 special zones and districts include Caofeidian and Baigouxincheng in Hebei Province, Yunzhou District and

Taigu District in Shanxi Province, Jiagedaqi in Heilongjiang Province, Shennongjia Forestry District in Hubei Province, Wangcheng District, Nanyue District and Datong Lake District in Hunan Province, Yangpu Development Zone in Hainan Province, Liuzhite District and shuicheng District in Guizhou Province, Kunming Yangzonghai Scenic Spot in Yunnan Province, Hongsibao Development Zone in Ningxia Autonomous Region, and Xihai Town and Dachaidan Administrative Commission in Haibei Prefecture of Qinghai Province.

(3) In the part of Villages and Towns, statistics is based on data from 18822 towns, 8876 townships, 447 special areas at the level of town and township, and 2.363 million natural villages including 493 thousand villages where villagers' committees are situated.

5. This Yearbook does not include data of Hong Kong Special Administrative Region, Macao Special Administrative Region and Taiwan Province.

6. All the per capita and coverage rate data in this Yearbook, except the per capita residential floor area and per capita daily consumption of domestic water, are calculated using the sum of resident and non-resident population as a denominator.

7. In this yearbook, "blank space" indicates that the figure is not large enough to be measured with the smallest unit in the table, or data are unknown or are not available.

8. The calculation errors of the total or relative value of some data in this Yearbook arising from the use of different measurement units have not been mechanically aligned.

9. Any comments to improve the quality of the yearbook are welcomed to promote the advancement in statistics in China's construction industry.

目　录

Contents

城市部分
Statistics for Cities

县城部分
Statistics for County Seats

一、综合数据
General Data

二、居民生活数据
Data by Residents Living

三、居民出行数据
Data by Residents Travel

四、环境卫生数据
Data by Environmental Health

五、绿色生态数据
Data by Green Ecology

村镇部分
Statistics for Villages and Small Towns

主要指标解释
Explanatory Notes on Main Indicators

城市部分

Statistics for Cities

一、综合数据
General Data

1-1-1　全国历年城市市政公用设施水平
Level of National Urban Service Facilities in Past Years

年份 Year	供水普及率 (%) Water Coverage Rate (%)	燃气普及率 (%) Gas Coverage Rate (%)	每万人拥有公共交通车辆 (标台) Motor Vehicle for Public Transport Per 10,000 Persons (standard unit)	人均道路面积 (平方米) Road Surface Area Per Capita (m²)	污水处理率 (%) Wastewater Treatment Rate (%)	园林绿化 Landscaping 人均公园绿地面积(平方米) Public Recreational Green Space Per Capita (m²)	建成区绿化覆盖率(%) Green Coverage Rate of Built District (%)	建成区绿地率(%) Green Space Rate of Built District (%)	每万人拥有公厕(座) Number of Public Lavatories per 10,000 Persons (unit)
1978									
1979									
1980									
1981	53.7	11.6		1.81		1.50			3.77
1982	56.7	12.6		1.96		1.65			3.99
1983	52.5	12.3		1.88		1.71			3.95
1984	49.5	13.0		1.84		1.62			3.57
1985	45.1	13.0		1.72		1.57			3.28
1986	51.3	15.2	2.5	3.05		1.84	16.90		3.61
1987	50.4	16.7	2.4	3.10		1.90	17.10		3.54
1988	47.6	16.5	2.2	3.10		1.76	17.00		3.14
1989	47.4	17.8	2.1	3.22		1.69	17.80		3.09
1990	48.0	19.1	2.2	3.13		1.78	19.20		2.97
1991	54.8	23.7	2.7	3.35	14.86	2.07	20.10		3.38
1992	56.2	26.3	3.0	3.59	17.29	2.13	21.00		3.09
1993	55.2	27.9	3.0	3.70	20.02	2.16	21.30		2.89
1994	56.0	30.4	3.0	3.84	17.10	2.29	22.10		2.69
1995	58.7	34.3	3.6	4.36	19.69	2.49	23.90		3.00
1996	60.7	38.2	3.8	4.96	23.62	2.76	24.43	19.05	3.02
1997	61.2	40.0	4.5	5.22	25.84	2.93	25.53	20.57	2.95
1998	61.9	41.8	4.6	5.51	29.56	3.22	26.56	21.81	2.89
1999	63.5	43.8	5.0	5.91	31.93	3.51	27.58	23.03	2.85
2000	63.9	45.4	5.3	6.13	34.25	3.69	28.15	23.67	2.74
2001	72.26	60.42	6.10	6.98	36.43	4.56	28.38	24.26	3.01
2002	77.85	67.17	6.73	7.87	39.97	5.36	29.75	25.80	3.15
2003	86.15	76.74	7.66	9.34	42.39	6.49	31.15	27.26	3.18
2004	88.85	81.53	8.41	10.34	45.67	7.39	31.66	27.72	3.21
2005	91.09	82.08	8.62	10.92	51.95	7.89	32.54	28.51	3.20
2006	86.07 (97.04)	79.11 (88.58)	9.05 (10.13)	11.04 (12.36)	55.67	8.30 (9.30)	35.11	30.92	2.88 (3.22)
2007	93.83	87.40	10.23	11.43	62.87	8.98	35.29	31.30	3.04
2008	94.73	89.55	11.13	12.21	70.16	9.71	37.37	33.29	3.12
2009	96.12	91.41		12.79	75.25	10.66	38.22	34.17	3.15
2010	96.68	92.04		13.21	82.31	11.18	38.62	34.47	3.02
2011	97.04	92.41		13.75	83.63	11.80	39.22	35.27	2.95
2012	97.16	93.15		14.39	87.30	12.26	39.59	35.72	2.89
2013	97.56	94.25		14.87	89.34	12.64	39.70	35.78	2.83
2014	97.64	94.57		15.34	90.18	13.08	40.22	36.29	2.79
2015	98.07	95.30		15.60	91.90	13.35	40.12	36.36	2.75
2016	98.42	95.75		15.80	93.44	13.70	40.30	36.43	2.72
2017	98.30	96.26		16.05	94.54	14.01	40.91	37.11	2.77
2018	98.36	96.70		16.70	95.49	14.11	41.11	37.34	2.88
2019	98.78	97.29		17.36	96.81	14.36	41.51	37.63	2.93
2020	98.99	97.87		18.04	97.53	14.78	42.06	38.24	3.07

注：1. 自2006年起，人均和普及率指标按城区人口和城区暂住人口合计为分母计算，以公安部门的户籍统计和暂住人口统计为准。括号中的数据为与往年同口径数据。

2. "人均公园绿地面积"指标2005年及以前年份为"人均公共绿地面积"。

3. 自2009年起，城市公共交通内容不再统计，增加城市轨道交通建设情况内容。

Note: 1. Since 2006, figure in terms of per capita and coverage rate have been calculated based on denominator which combines both permanent and temporary residents in urban areas. And the population should come from statistics of police. The data in brackets are same index calculated by the method of past years.

2. Since 2005, Public Green Space Per Capita is changed to Public Recreational Green Space Per Capita.

3. Since 2009, statistics on urban public transport have been removed, and relevant information on the construction of rail transit system has been added.

1-1-2　全国城市市政公用设施水平(2020年)

地区名称 Name of Regions	人口密度 (人/ 平方公里) Population Density (person/ square kilometer)	人均日生 活用水量 (升) Daily Water Consumption Per Capita (liter)	供　水 普及率 (%) Water Coverage Rate (%)	公共供水 普及率 Public Water Coverage Rate	燃　气 普及率 (%) Gas Coverage Rate (%)	建成区供水 管道密度 (公里/ 平方公里) Density of Water Supply Pipelines in Built District (kilometer/ square kilometer)	人均道路 面　积 (平方米) Road Surface Area Per Capita (m²)	建成区 路网密度 (公里/ 平方公里) Road in Built District (kilometer/ square kilometer)
全　国 National Total	**2778**	**179.40**	**98.99**	**97.84**	**97.87**	**14.02**	**18.04**	**7.07**
北　京 Beijing		154.19	98.39	92.26	100.00		7.67	
天　津 Tianjin	4449	115.69	100.00	100.00	100.00	17.90	14.91	6.77
河　北 Hebei	3085	127.34	100.00	98.61	99.72	9.93	21.06	8.14
山　西 Shanxi	4015	133.86	99.60	96.98	98.70	9.76	18.41	7.20
内　蒙　古 Inner Mongolia	1850	101.25	99.50	97.32	97.26	9.50	23.93	7.65
辽　宁 Liaoning	1805	149.98	99.71	98.41	98.75	13.15	16.21	6.78
吉　林 Jilin	1876	121.79	95.60	93.78	92.93	9.96	15.71	6.02
黑　龙　江 Heilongjiang	5501	129.47	98.99	97.48	90.82	11.98	15.59	7.11
上　海 Shanghai	3830	203.92	100.00	100.00	100.00	31.95	4.76	4.47
江　苏 Jiangsu	2240	220.69	100.00	99.64	99.92	18.33	25.60	8.91
浙　江 Zhejiang	2105	220.06	100.00	99.55	100.00	20.46	19.08	7.38
安　徽 Anhui	2655	197.37	99.60	98.96	99.24	14.67	24.29	6.99
福　建 Fujian	3545	214.47	99.90	99.89	99.21	18.98	18.83	7.87
江　西 Jiangxi	4426	176.68	98.62	98.49	97.59	14.74	19.81	6.91
山　东 Shandong	1665	119.35	99.84	97.69	99.31	9.99	25.64	7.71
河　南 Henan	4994	128.99	98.19	95.52	96.83	8.90	15.32	5.11
湖　北 Hubei	2778	192.85	99.56	98.76	98.40	15.34	18.89	8.10
湖　南 Hunan	3677	211.47	98.94	98.66	97.29	15.26	19.72	6.93
广　东 Guangdong	3909	236.78	98.49	98.39	98.99	17.24	13.26	6.25
广　西 Guangxi	2162	263.48	99.68	99.17	99.36	14.31	23.76	8.35
海　南 Hainan	2444	275.27	98.02	96.77	98.95	2.54	17.91	11.39
重　庆 Chongqing	2070	179.80	94.69	93.62	96.16	15.83	14.65	6.56
四　川 Sichuan	3158	196.39	98.28	98.09	97.41	15.09	18.13	7.45
贵　州 Guizhou	2262	169.78	98.90	98.90	94.55	8.70	21.23	7.11
云　南 Yunnan	3138	155.13	98.10	97.43	78.65	10.87	16.62	6.29
西　藏 Tibet	1584	290.86	98.78	98.68	63.20	10.36	20.74	4.01
陕　西 Shaanxi	4985	155.71	97.88	93.19	98.62	8.16	16.73	5.18
甘　肃 Gansu	3235	139.98	98.15	96.12	94.80	6.88	20.25	6.42
青　海 Qinghai	2930	138.19	98.69	97.36	93.81	12.94	18.91	6.04
宁　夏 Ningxia	3153	162.57	98.87	97.73	97.52	5.96	26.78	5.66
新　疆 Xinjiang	4036	163.09	99.51	99.24	98.64	8.68	25.36	5.80
新疆生产 建设兵团 Xinjiang Production and Construction Corps	2063	169.75	99.61	99.61	96.55	9.52	24.84	6.82

注：本表中北京市的建成区绿化覆盖率和建成区绿地率均为该市调查面积内数据，全国城市建成区绿化覆盖率和建成区绿
　　地率作适当修正。

National Level of National Urban Service Facilities(2020)

建成区道路面积率(%) Road Surface Area Rate of Built District (%)	建成区排水管道密度(公里/平方公里) Density of Sewers in Built District (kilometer/square kilometer)	污水处理率(%) Wastewater Treatment Rate (%)	污水处理厂集中处理率 Centralized Treatment Rate of Wastewater Treatment Plants	人均公园绿地面积(平方米) Public Recreational Green Space Per Capita (m²)	建成区绿化覆盖率(%) Green Coverage Rate of Built District (%)	建成区绿地率(%) Green Space Rate of Built District (%)	生活垃圾处理率(%) Domestic Garbage Treatment Rate (%)	生活垃圾无害化处理率 Domestic Garbage Harmless Treatment Rate	地区名称 Name of Regions
14.19	11.11	97.53	95.78	14.78	42.06	38.24	99.92	99.75	全　国
		96.56	94.76	16.59	48.96	46.69	100.00	100.00	北　京
12.60	18.72	96.42	95.54	10.31	37.59	34.46	100.00	100.00	天　津
17.52	9.04	98.46	98.46	15.30	42.92	39.26	100.00	100.00	河　北
15.73	7.68	99.60	99.60	13.51	43.88	39.93	100.00	100.00	山　西
17.09	10.11	97.80	97.80	19.20	40.45	37.39	99.92	99.92	内　蒙　古
11.98	7.43	98.23	97.44	13.40	41.73	39.22	100.00	99.53	辽　宁
10.90	6.97	97.69	97.69	12.94	40.40	35.68	100.00	100.00	吉　林
11.52	7.02	95.98	92.16	12.77	36.88	33.15	99.87	99.87	黑　龙　江
9.33	13.92	96.68	96.17	9.05	37.32	35.79	100.00	100.00	上　海
15.43	14.86	96.82	90.19	15.34	43.47	40.11	100.00	100.00	江　苏
14.81	13.88	97.69	96.75	13.59	42.22	38.07	100.00	100.00	浙　江
17.05	13.49	97.43	95.76	14.88	42.01	38.49	100.00	100.00	安　徽
13.71	10.94	97.15	93.40	14.94	44.63	40.94	100.00	100.00	福　建
13.86	11.00	97.48	96.47	14.80	46.35	42.72	100.00	100.00	江　西
16.06	11.32	98.26	98.11	17.68	41.65	37.76	100.00	100.00	山　东
12.79	8.86	98.32	98.30	14.43	41.92	36.48	99.94	99.94	河　南
15.26	10.29	96.97	92.24	13.83	41.07	36.55	100.00	100.00	湖　北
16.12	9.99	97.79	96.95	12.16	41.51	37.20	100.00	100.00	湖　南
11.48	13.95	97.66	97.57	18.14	43.49	39.39	99.95	99.95	广　东
16.47	11.62	98.99	88.73	12.85	41.30	35.94	100.00	100.00	广　西
11.93	10.28	98.68	98.53	11.62	40.62	36.81	100.00	100.00	海　南
14.22	13.94	98.17	97.84	16.50	43.05	39.62	99.46	93.84	重　庆
15.21	10.96	96.86	93.57	14.40	42.48	37.39	99.99	99.99	四　川
13.85	8.28	97.44	97.44	17.04	40.94	38.82	97.85	97.85	贵　州
13.13	11.25	97.63	96.65	12.27	40.46	36.57	99.99	99.99	云　南
9.60	3.38	96.28	96.28	12.02	38.06	35.93	99.63	99.63	西　藏
14.26	7.34	96.79	96.79	12.79	40.80	37.09	99.93	99.93	陕　西
13.72	7.97	97.18	97.18	15.15	36.28	32.48	100.00	100.00	甘　肃
15.97	13.46	95.31	95.31	12.45	35.90	33.71	99.29	99.29	青　海
15.21	4.38	96.74	96.74	21.02	41.95	39.65	99.96	99.96	宁　夏
12.71	5.71	98.11	98.11	13.19	40.42	36.37	99.12	99.12	新　疆
11.84	6.10	99.44	99.44	20.18	43.68	40.87	99.63	98.58	新疆生产建设兵团

Note: All of the green coverage rate and green space rate for the built-up areas of Beijing Municipality in the Table refer to the data for the areas surveyed in the city. The green coverage rate and green space rate for the nationwide urban built-up areas have been revised appropriately.

1-2-1 全国历年城市数量及人口、面积情况

面积计量单位: 平方公里
人口计量单位: 万人

年 份 Year	城市个数 Number of Cities	地级 City at Prefecture Level	县级 City at County Level	县及其他 个 数 Number of Counties	城区人口 Urban Population
1978	193	98	92	2153	7682.0
1979	216	104	109	2153	8451.0
1980	223	107	113	2151	8940.5
1981	226	110	113	2144	14400.5
1982	245	109	133	2140	14281.6
1983	281	137	141	2091	15940.5
1984	300	148	149	2069	17969.1
1985	324	162	159	2046	20893.4
1986	353	166	184	2017	22906.2
1987	381	170	208	1986	25155.7
1988	434	183	248	1936	29545.2
1989	450	185	262	1919	31205.4
1990	467	185	279	1903	32530.2
1991	479	187	289	1894	29589.3
1992	517	191	323	1848	30748.2
1993	570	196	371	1795	33780.9
1994	622	206	413	1735	35833.9
1995	640	210	427	1716	37789.9
1996	666	218	445	1696	36234.5
1997	668	222	442	1693	36836.9
1998	668	227	437	1689	37411.8
1999	667	236	427	1682	37590.0
2000	663	259	400	1674	38823.7
2001	662	265	393	1660	35747.3
2002	660	275	381	1649	35219.6
2003	660	282	374	1642	33805.0
2004	661	283	374	1636	34147.4
2005	661	283	374	1636	35923.7
2006	656	283	369	1635	33288.7
2007	655	283	368	1635	33577.0
2008	655	283	368	1635	33471.1
2009	654	283	367	1636	34068.9
2010	657	283	370	1633	35373.5
2011	657	284	369	1627	35425.6
2012	657	285	368	1624	36989.7
2013	658	286	368	1613	37697.1
2014	653	288	361	1596	38576.5
2015	656	291	361	1568	39437.8
2016	657	293	360	1537	40299.2
2017	661	294	363	1526	40975.7
2018	673	302	371	1519	42730.0
2019	679	300	379	1516	43503.7
2020	687	301	386	1495	44253.7

注: 1.2005年及以前年份"城区人口"为"城市人口", "城区面积"为"城市面积"。
2.2005年、2009年、2011年城市建设用地面积不含上海市。2020年城区面积、建成区面积和城市建设用地面积不含
北京市。

National Changes in Number of Cities, Urban Population and Urban Area in Past Years

Area Measurement Unit: Square Kilometer
Population Measurement Unit: 10,000 persons

非农业 人　口 Non-Agricultural Population	城区暂住 人　口 Urban Temporary Population	城区面积 Urban Area	建成区 面　积 Area of Built District	城市建设 用地面积 Area of Urban Construction Land	年　份 Year
					1978
					1979
					1980
9243.6		206684.0	7438.0	6720.0	1981
9590.0		335382.3	7862.1	7150.5	1982
10047.2		366315.9	8156.3	7365.6	1983
10956.9		480733.3	9249.0	8480.4	1984
11751.3		458066.2	9386.2	8578.6	1985
12233.8		805834.0	10127.3	9201.6	1986
12893.1		898208.0	10816.5	9787.9	1987
13969.5		1052374.2	12094.6	10821.6	1988
14377.7		1137643.5	12462.2	11170.7	1989
14752.1		1165970.0	12855.7	11608.3	1990
14921.0		980685.0	14011.1	12907.9	1991
15459.4		969728.0	14958.7	13918.1	1992
16550.1		1038910.0	16588.3	15429.8	1993
17665.5		1104712.0	17939.5	20796.2	1994
18490.0		1171698.0	19264.2	22064.0	1995
18882.9		987077.9	20214.2	19001.6	1996
19469.9		835771.8	20791.3	19504.6	1997
19861.8		813585.7	21379.6	20507.6	1998
20161.6		812817.6	21524.5	20877.0	1999
20952.5		878015.0	22439.3	22113.7	2000
21545.5		607644.3	24026.6	24192.7	2001
22021.2		467369.3	25972.6	26832.6	2002
22986.8		399173.2	28308.0	28971.9	2003
23635.9		394672.5	30406.2	30781.3	2004
23652.0		412819.1	32520.7	29636.8	2005
	3984.1	166533.5	33659.8	34166.7	2006
	3474.3	176065.5	35469.7	36351.7	2007
	3517.2	178110.3	36295.3	39140.5	2008
	3605.4	175463.6	38107.3	38726.9	2009
	4095.3	178691.7	40058.0	39758.4	2010
	5476.8	183618.0	43603.2	41805.3	2011
	5237.1	183039.4	45565.8	45750.7	2012
	5621.1	183416.1	47855.3	47108.5	2013
	5951.5	184098.6	49772.6	49982.7	2014
	6561.5	191775.5	52102.3	51584.1	2015
	7414.0	198178.6	54331.5	52761.3	2016
	8164.1	198357.2	56225.4	55155.5	2017
	8421.7	200896.5	58455.7	56075.9	2018
	8911.9	200569.5	60312.5	58307.7	2019
	9509.0	186628.9	60721.3	58355.3	2020

Note: 1.In 2005 and before, Urban Population is the population of the city proper, and Urban Area is the area of the city proper.

2.Urban usable land for construction purpose throughout the country does not include that in Shanghai in 2005 ,2009 and 2011. Urban Area, Area of Built District and Urban usable land for construction purpose throughout the country does not include that in Beijing in 2020.

1-2-2　全国城市人口和建设用地(2020年)

面积计量单位: 平方公里

人口计量单位: 万人

地区名称 Name of Regions	市区面积 Urban District Area	市区人口 Urban District Population	市区暂住人口 Urban District Temporary Population	城区面积 Urban Area	城区人口 Urban Population	城区暂住人口 Urban Temporary Population	建成区面积 Area of Built District	小计 Subtotal
全　国　National Total	2322592	81643	12983	186628.87	44253.74	9509.02	60721.32	58355.29
北　京　Beijing		2189			1916.40			
天　津　Tianjin	11760	1387		2639.78	1174.44		1170.24	1040.63
河　北　Hebei	49498	3722	211	6320.88	1796.20	154.10	2236.46	2067.10
山　西　Shanxi	32068	1685	230	3020.04	1061.08	151.41	1267.19	1239.13
内 蒙 古　Inner Mongolia	148743	937	277	4984.11	686.88	235.39	1262.18	1152.66
辽　宁　Liaoning	75477	3085	363	12509.14	1980.46	277.71	2725.60	2798.70
吉　林　Jilin	109277	1920	205	6485.80	1023.10	193.37	1565.85	1504.69
黑 龙 江　Heilongjiang	220375	2297	195	2573.80	1237.09	178.67	1826.88	1799.98
上　海　Shanghai	6341	2428		6340.50	2428.14		1237.85	1944.96
江　苏　Jiangsu	69961	5889	1302	15796.74	3052.17	485.85	4786.78	4681.89
浙　江　Zhejiang	55869	3655	1872	13461.43	1816.34	1016.68	3157.16	3054.63
安　徽　Anhui	46905	2873	616	6712.30	1314.27	467.97	2409.89	2321.27
福　建　Fujian	45738	2227	818	3919.08	1008.90	380.56	1648.14	1589.46
江　西　Jiangxi	46878	2258	206	2996.54	1161.70	164.62	1703.63	1603.41
山　东　Shandong	92953	6803		23953.82	3396.29	592.85	5646.34	5102.39
河　南　Henan	47286	4558	541	5364.36	2201.52	477.66	3039.98	2882.19
湖　北　Hubei	97662	4259	483	8220.53	1900.37	383.20	2756.81	2665.89
湖　南　Hunan	54248	3023	318	4778.54	1520.45	236.52	1959.38	1849.50
广　东　Guangdong	99927	8280	2680	16213.21	4392.14	1944.93	6501.44	5892.48
广　西　Guangxi	75177	2519	311	5876.81	985.10	285.47	1617.61	1572.29
海　南　Hainan	17065	590	169	1439.41	225.86	125.99	390.70	373.04
重　庆　Chongqing	43264	2602	500	7779.14	1213.56	396.62	1565.61	1452.62
四　川　Sichuan	86811	4341	620	8894.04	2325.10	483.22	3190.49	2993.18
贵　州　Guizhou	35644	1438	173	3702.30	680.02	157.32	1118.37	1019.69
云　南　Yunnan	88367	1744	175	3274.28	905.58	121.77	1266.24	1238.27
西　藏　Tibet	48869	92	43	632.17	57.61	42.55	168.39	157.51
陕　西　Shaanxi	55458	2040	145	2597.11	1189.85	104.76	1372.20	1346.35
甘　肃　Gansu	88553	985	137	2004.76	536.78	111.66	901.43	919.26
青　海　Qinghai	203480	301	24	735.99	191.06	24.55	235.20	224.67
宁　夏　Ningxia	21889	434	63	951.27	246.37	53.53	494.26	451.60
新　疆　Xinjiang	233664	959	274	1943.17	556.89	227.37	1300.15	1224.15
新疆生产建设兵团　Xinjiang Production and Construction Corps	13384	124	34	507.82	72.02	32.72	198.87	191.70

注: 1.本表中山东省市区人口为常住人口。

Note:1.The Urban District Population of Shandong Province in this table refers to the data of the resident population.

National Urban Population and Construction Land (2020)

Area Measurement Unit: Square Kilometer
Population Measurement Unit: 10,000persons

城市建设用地面积　Area of Urban Construction Land								本年征用土地面积	耕地	地区名称
居住用地 Residential	公共管理与公共服务用地 Administration and Public Services	商业服务业设施用地 Commercial and Business Facilities	工业用地 Industrial, Manufacturing	物流仓储用地 Logistics and Warehouse	道路交通设施用地 Road, Street and Transportation	公用设施用地 Municipal Utilities	绿地与广场用地 Green Space and Square	Area of Land Requisition This Year	Arable Land	Name of Regions
18098.71	5162.30	4088.64	11339.88	1586.19	9503.42	1758.08	6818.07	2362.38	999.14	全　国
										北　京
289.37	81.36	86.80	238.79	51.19	165.52	20.36	107.24	19.62	9.65	天　津
639.42	155.68	126.97	234.12	53.72	353.26	57.49	446.44	34.60	23.47	河　北
409.69	132.19	91.69	180.84	46.92	218.81	29.13	129.86	495.50	161.59	山　西
362.10	103.27	97.71	132.20	46.14	222.11	37.03	152.10	17.25	3.53	内　蒙　古
919.44	174.57	189.29	697.79	86.50	411.63	71.65	247.83	31.37	18.80	辽　宁
502.71	114.40	95.25	307.68	50.54	228.39	55.48	150.24	40.75	25.60	吉　林
632.96	155.51	97.22	361.91	70.41	267.89	48.78	165.30	17.36	9.62	黑　龙　江
549.77	153.19	124.66	537.74	52.31	139.05	247.48	140.76	23.39	13.49	上　海
1344.88	361.54	366.47	1016.18	108.54	786.70	105.78	591.80	152.67	70.33	江　苏
942.88	259.09	210.79	699.93	60.57	533.26	70.60	277.51	137.82	77.96	浙　江
685.57	174.80	162.50	451.72	45.84	404.94	64.74	331.16	131.53	74.82	安　徽
522.72	166.71	108.30	266.15	32.20	281.66	43.26	168.46	56.65	12.49	福　建
484.43	168.34	106.06	304.89	32.91	253.83	40.59	212.36	83.01	29.34	江　西
1624.33	473.75	389.02	1094.46	144.13	724.45	121.54	530.71	150.03	74.47	山　东
875.74	303.15	146.85	348.68	79.53	465.79	104.19	558.26	25.04	15.64	河　南
810.40	247.06	159.96	632.39	79.33	420.21	96.16	220.38	157.36	68.65	湖　北
700.48	215.69	130.83	273.47	41.31	247.08	73.27	167.37	66.12	19.32	湖　南
1793.75	447.39	399.36	1551.57	108.33	1125.41	102.64	364.03	148.09	47.39	广　东
455.84	142.71	94.03	214.80	50.16	289.25	58.22	267.28	79.88	27.77	广　西
134.02	52.20	42.69	22.13	3.99	69.69	4.03	44.29	17.33	1.80	海　南
417.60	139.10	86.27	299.12	31.41	313.19	31.72	134.21	115.50	61.01	重　庆
936.43	260.19	230.74	497.94	81.91	510.65	68.36	406.96	179.70	73.01	四　川
319.35	117.05	89.05	170.43	25.09	172.37	20.84	105.51	14.30	3.86	贵　州
428.22	135.70	101.73	130.52	35.91	186.77	34.90	184.52	36.71	13.95	云　南
35.38	23.69	15.29	15.67	4.12	24.23	9.68	29.45	9.00	2.31	西　藏
428.26	133.91	103.73	175.04	33.30	213.33	26.52	232.26	68.65	32.96	陕　西
232.51	83.34	67.90	205.32	41.70	133.55	35.99	118.95	16.52	10.31	甘　肃
66.01	17.17	13.70	16.49	17.17	40.06	10.81	43.26	1.33	0.07	青　海
134.07	42.41	35.47	45.07	10.32	84.67	10.81	88.78	11.75	8.29	宁　夏
366.70	107.60	101.27	189.30	56.11	182.85	50.91	169.41	16.60	3.98	新　疆
53.68	19.54	17.04	27.54	4.58	32.82	5.12	31.38	6.95	3.66	新疆生产建设兵团

1-3-1　全国历年城市维护建设资金收入

计量单位：万元

年份 Year	合计 Total	城市维护 建 设 税 Urban Maintenance and Construction Tax	城市公用 事业附加 Extra- Charges for Municipal Utilities	中央财政 拨　　款 Financial Allocation From Central Government Budget	地方财政 拨　　款 Financial Allocation From Local Government Budget
1978					
1979					
1980	276174		92966	66212	
1981	345600		92186	65657	
1982	425957		97416	69381	
1983	408524		103142	81571	
1984	468835		118445	96120	
1985	1168293	331296	129744	143313	123963
1986	1448311	406817	157889	138944	351961
1987	1638197	442275	176337	152094	225498
1988	1845193	525452	192643	104666	181053
1989	1835625	603583	198409	91529	170347
1990	2104896	650908	226186	109126	198421
1991	2661198	695681	270109	99919	277595
1992	3934788	779701	310916	134237	576569
1993	5811904	980234	329620	269593	594615
1994	6748008	1161193	413851	213565	599378
1995	7743732	1409097	447388	213543	725758
1996	8476420	1578088	555873	103699	862641
1997	11103424	1906696	561217	139392	1154245
1998	14333158	2153203	603965	649420	1581006
1999	16271209	2191985	631558	1052000	1714628
2000	19889324	2372908	541475	1150668	2081330
2001	25262680	2709430	488144	895818	3237790
2002	31561758	3160358	498761	759545	3927309
2003	42761892	3717417	556909	771320	5329302
2004	52575966	4462912	590251	526414	6657678
2005	54225147	5512933	554799	621597	7958871
2006	35406259	5667745	762171	566632	10748459
2007	47617452	6170604	824251	348134	12137926
2008	56164219	7442775	896248	756011	14247316
2009	67276878	7719472	980433	1066464	22311926
2010	85704996	9970340	1090649	1747961	16855095
2011	117817189	13395522	1571916	1383183	19817577
2012	119233467	14797543	1663517	2453300	27252469
2013	143227465	17070039	1874070		
2014	138005449	17843612	1819986		
2015	160735509	19421884	2255346		
2016	180194600	22429127	2100389		

注：1.自2006年起，城市维护建设资金收入仅包含财政性资金，不含社会融资。地方财政拨款中包括省、市财政专项拨款
　　和市级以下财政资金；其他收入中包括市政公用设施配套费、市政公用设施有偿使用费、土地出让转让金、资产置
　　换收入及其他财政性资金。
　　2.2014年数据不含北京市。

National Revenue of Urban Maintenance and Construction Fund in Past Years

Measurement Unit: 10,000 RMB

水资源费 Water Resource Fee	国内贷款 Domestic Loan	利用外资 Foreign Investment	企事业单位自筹资金 Self-Raised Funds by Enterprises and Institutions	其他收入 Other Revenues	年份 Year
					1978
					1979
				116996	1980
				187757	1981
				259160	1982
				223811	1983
				254270	1984
14142			49362	376473	1985
16812	31716	744		343428	1986
22420	61601	748		557224	1987
24396	75829	6386		734768	1988
24873	44494	10078		692312	1989
27876	88429	24713		779237	1990
35324	229274	107127		946169	1991
42506	322596	73484		1694779	1992
48498	445658	138159		3005527	1993
47060	413327	183295		3716339	1994
51964	476654	254876		4164452	1995
60509	956850	558688	1194722	2605348	1996
61203	1657405	1407194	1077986	3138088	1997
63724	3069622	738402	1769411	3704405	1998
77828	3741979	494463	2374802	3991966	1999
100590	4146988	847124	3332242	5315999	2000
111087	7416589	563167	4095561	5745094	2001
123825	8739016	610535	6007620	7734789	2002
159757	13318273	681057	7696801	10531056	2003
206883	14455465	741815	9002038	15932510	2004
249990	16698914	927141	9460261	12240641	2005
244522				17416730	2006
279545				27856992	2007
254199				32567671	2008
247971				34950612	2009
295347				55845604	2010
750347				80898644	2011
413782				72652856	2012
					2013
					2014
					2015
					2016

Note: 1.Since 2006, national revenue of urban maintenance and construction fund includes the fund fiscal budget, not including social funds. Local Financail Allocation include province and city special financial allocation, and Other Revenues include fees for expansion of municipal utilites capacity, fees for use of municipal utilities, land transfer income, asset replacement income and other financial funds.

2.The data of 2014 did not include those of Beijing Municipality.

1-3-2　全国历年城市维护建设资金支出

计量单位: 万元

| 年 份
Year | 支出合计
Total | 按用途分　By Purpose | | | 供 水
Water Supply | 燃 气
Gas Supply | 集中供热
Central Heating |
		维护支出 Maintenance Expenditure	固定资产投资支出 Expenditure from Investment in Fixed Assets	其他支出 Other Expenditures			
1978	163589		163589		38439		
1979	256895	142593	114302		49317		
1980	266451	152991	112370	1090	45258		
1981	325615	180778	144837		32932		
1982	369382	172902	183897	12583	50238		
1983	426986	214709	185511	26766	51642		
1984	462705	219226	243479		61886		
1985	1122714	375563	721943	25208	102424	86598	12157
1986	1343025	431595	816154	95276	139303	110697	17993
1987	1480706	479768	895695	105243	50997	99993	20339
1988	644695	303420	246835	94440	78875		
1989	737038	361213	249516	126309	99822		
1990	814901	406025	269992	138884	108132		
1991	2442190	446562	1800536	195092	265064	224584	
1992	3841602	1086759	2529896	224947	453996	261810	
1993	5573279	1550715	3757865	264699	613695	394564	
1994	6583051	1721457	4514337	347257	753151	376545	
1995	7578470	1092891	6056988	428591	839579	332546	
1996	8827009	2029361	6272955	524693	991630	405386	149145
1997	11229697	2149827	7832004	1247866	915220	477144	255485
1998	14374965	2354278	10927777	1092910	1250356	552624	406733
1999	16172386	2600945	12281137	1290304	1175347	485483	425240
2000	18960338	2699820	15400602	859916	1232155	620982	638120
2001	25372970	2367551	19207735	3797684	1530995	775355	789525
2002	31780253	3184473	24475402	4120378	1578903	844230	1137775
2003	42479605	3840924	32807384	5831298	1769459	1120184	1438295
2004	46617124	4662527	36924771	5029826	2013230	1237042	1538473
2005	52757240	5474379	40946915	6335946	2140834	1210925	1838492
2006	33494964	6950298	20005661	6888341	791690	344728	583932
2007	42473049	8689485	24649406	9134158	981690	409311	617348
2008	50083394	10432730	30612349	9279334	1074319	565441	1047382
2009	59270667	10982852	38125424	10187001	1175896	512700	1206084
2010	75080799	13443275	47144531	14389306	1751365	823824	1450987
2011	87390666	16790270	53139075	17461321			
2012	101981275	20228860	63505657	18246758			
2013	108047393						
2014	106589135						
2015	124386269						
2016	138326496						

注: 1.1978年至1984年, 1988年至1990年供水支出为公用事业支出; 1978年至2000年道路桥梁支出为市政工程支出, 包含道路、桥梁、排水等。

2.自2006年起, 城市维护建设资金支出仅包含财政性资金, 不含社会融资。

3."公共交通"中, 2009年至2010年数据仅包括城市轨道交通建设投资支出。

4.2014年数据不含北京市。

National Expenditure of Urban Maintenance and Construction Fund in Past Years

Measurement Unit: 10,000 RMB

按行业分　By Indurstry							
公共交通 Public Transportation	道路桥梁 Road and Bridge	排　水 Sewerage	防　洪 Flood Control	园林绿化 Landscaping	市容环境 卫　生 Environmental Sanitation	其　他 Other	年　份 Year
	48882			13904			1978
	67940			19337	26541		1979
	77891			24708	26093	1089	1980
	65032			23311	31751		1981
	114715			30304	41112	12585	1982
	121204			35445	44873	26766	1983
	150554			44421	52459	-6799	1984
56073	331930			85831	76250	25208	1985
43281	408526	86533		93234	107719	95276	1986
50997	481682	92141		98347	109356	105242	1987
	214471			61143	89510	94440	1988
	229082			66320	94189	126309	1989
	256539			73356	114880	138884	1990
69668	757003			186574	206776	195092	1991
106503	1198511			245334	283664	224947	1992
199547	2336633			311418	363825	264699	1993
139599	2929744			402105	440382	347257	1994
154127	3222256			523299	575786	428591	1995
201337	3590192	393753		532873	616247	524693	1996
285625	4664378	539945		675787	736913	1247866	1997
509899	6263259	881923		1035178	990036	1092909	1998
504331	7359954	985336		1315396	953377	1290304	1999
1434521	8188958	1291761		1779059	1526633	859916	2000
1901819	8640772	2181103	568835	1776331	932631	6275604	2001
2714424	11081731	2670957	1394590	918210	2628912	6810521	2002
2540364	17463295	3580353	1089181	3348090	1665167	8465217	2003
2405912	19862495	3361557	1494379	953154	3619398	10131484	2004
3467801	23719550	3555419	1786861	908127	4224763	9904468	2005
1511917	13154564	2215124	448841	2967691	1746169	9730308	2006
3394326	16951020	2972994	729391	3611468	2100669	10704832	2007
4012497	19769532	3709526	977739	4086639	2579521	12260798	2008
2259540	23061301	4923007	820261	4979200	2716274	17616404	2009
2530155	30396564	5808060	1604103	6908970	3666153	20140618	2010
							2011
							2012
							2013
							2014
							2015
							2016

Note: 1.Data in the water supply column from 1978 to 1984 and from 1988 to 1990 reder to expenditure of public utilities; Data in the bridge and road column refer to expenditure of municipal engineering construction projects,including roads, bridges and sewers.

2.From 2006, expenditure of Urban maintenance and construction fund includes the fund from fiscal buget, not including society fund.

3.In the section of "Public Transit", data from 2009 to 2010 only included investment in the construction of urban rail infrastructure.

4.The data of 2014 did not include those of Beijing Municipality.

1-4-1　按行业分全国历年城市市政公用设施建设固定资产投资

计量单位: 亿元

年 份 Year	本 年 固 定 资 产 投资总额 Completed Investment of This Year	供　水 Water Supply	燃　气 Gas Supply	集中供热 Central Heating	轨道交通 Rail Transit System	道路桥梁 Road and Bridge	排　水 Sewerage
1978	12.0	4.7				2.9	
1979	14.2	3.4	0.6		1.8	3.1	1.2
1980	14.4	6.7				7.0	
1981	19.5	4.2	1.8		2.6	4.0	2.0
1982	27.2	5.6	2.0		3.1	5.4	2.8
1983	28.2	5.2	3.2		2.8	6.5	3.3
1984	41.7	6.3	4.8		4.7	12.2	4.3
1985	64.0	8.1	8.2		6.0	18.6	5.6
1986	80.1	14.3	12.5	1.6	5.6	20.5	6.0
1987	90.3	17.2	10.9	2.1	5.5	27.1	8.8
1988	113.2	23.1	11.2	2.8	6.0	35.6	10.0
1989	107.0	22.2	12.3	3.3	7.7	30.1	9.7
1990	121.2	24.8	19.4	4.5	9.1	31.3	9.6
1991	170.9	30.2	24.8	6.4	9.8	51.8	16.1
1992	283.2	47.7	25.9	11.0	14.9	90.6	20.9
1993	521.8	69.9	34.8	10.7	22.1	191.8	37.0
1994	666.0	90.3	32.5	13.4	25.1	279.8	38.3
1995	807.6	112.4	32.9	13.8	30.9	291.6	48.0
1996	948.6	126.1	48.3	15.7	38.8	354.2	66.8
1997	1142.7	128.3	76.0	25.1	43.2	432.4	90.1
1998	1477.6	161.0	82.0	37.3	86.1	616.2	154.5
1999	1590.8	146.7	72.1	53.6	103.1	660.1	142.0
2000	1890.7	142.4	70.9	67.8	155.7	737.7	149.3
2001	2351.9	169.4	75.5	82.0	194.9	856.4	224.5
2002	3123.2	170.9	88.4	121.4	293.8	1182.2	275.0
2003	4462.4	181.8	133.5	145.8	281.9	2041.4	375.2
2004	4762.2	225.1	148.3	173.4	328.5	2128.7	352.3
2005	5602.2	225.6	142.4	220.2	476.7	2543.2	368.0
2006	5765.1	205.1	155.0	223.6	604.0	2999.9	331.5
2007	6418.9	233.0	160.1	230.0	852.4	2989.0	410.0
2008	7368.2	295.4	163.5	269.7	1037.2	3584.1	496.0
2009	10641.5	368.8	182.2	368.7	1737.6	4950.6	729.8
2010	13363.9	426.8	290.8	433.2	1812.6	6695.7	901.6
2011	13934.2	431.8	331.4	437.6	1937.1	7079.1	770.1
2012	15296.4	410.4	414.5	630.3	2064.5	7402.5	704.5
2013	16350.0	524.7	425.6	596.0	2455.1	8355.6	778.9
2014	16245.0	475.3	416.0	575.4	3221.2	7643.9	900.0
2015	16204.4	619.9	350.5	516.8	3707.1	7414.0	982.7
2016	17460.0	545.8	408.9	481.9	4079.5	7564.3	1222.5
2017	19327.6	580.1	445.7	584.2	5045.2	6996.7	1343.6
2018	20123.2	543.0	295.1	420.0	6046.9	6922.4	1529.9
2019	20126.3	560.1	242.7	333.0	5855.6	7655.3	1562.4
2020	22283.9	749.4	238.6	393.8	6420.8	7814.3	2114.8

注: 1.2008年及以前年份，"轨道交通"投资为"公共交通"，2009年以后仅包含轨道交通建设投资。

　　2.自2013年开始，全国城市市政公用设施建设固定资产投资中不再包括城市防洪固定资产投资。

National Fixed Assets Investment in Urban Service Facilities by Industry in Past Years

Measurement Unit: 100 million RMB

污水处理及其再生利用 Wastewater Treatment and reused	防洪 Flood Control	园林绿化 Land-scaping	市容环境卫生 Environmental Sanitation	垃圾处理 Garbage Treatment	地下综合管廊 Utility Tunnel	其他 Other	年份 Year
						4.4	1978
	0.1	0.4	0.1			3.4	1979
						0.7	1980
	0.2	0.9	0.7			3.2	1981
	0.3	1.1	0.9			5.9	1982
	0.4	1.2	0.9			4.7	1983
	0.5	2.0	0.9			5.9	1984
	0.9	3.3	2.0			11.3	1985
	1.6	3.4	2.8			11.9	1986
	1.4	3.4	2.1			11.9	1987
	1.6	3.4	2.6			16.9	1988
	1.2	2.8	2.8			14.8	1989
	1.3	2.9	2.9			15.4	1990
	2.1	4.9	3.6			21.3	1991
	2.9	7.2	6.5			55.6	1992
	5.9	13.2	10.6			125.8	1993
	8.0	18.2	10.9			149.7	1994
	9.5	22.5	13.6			232.5	1995
	9.1	27.5	12.5			249.7	1996
	15.5	45.1	20.9			266.1	1997
	35.8	78.4	36.7			189.6	1998
	43.0	107.1	37.1			226.0	1999
	41.9	143.2	84.3			297.5	2000
116.4	70.5	163.2	50.6	23.5		466.6	2001
144.1	135.1	239.5	64.8	29.7		551.0	2002
198.8	124.5	321.9	96.0	35.3		760.4	2003
174.5	100.3	359.5	107.8	53.0		838.4	2004
191.4	120.0	411.3	147.8	56.7		947.0	2005
151.7	87.1	429.0	175.8	51.8		554.3	2006
212.2	141.4	525.6	141.8	53.0		735.6	2007
264.7	119.6	649.8	222.0	50.6		530.8	2008
418.6	148.6	914.9	316.5	84.6		923.9	2009
521.4	194.4	1355.1	301.6	127.4		952.2	2010
420.5	243.8	1546.2	384.1	199.2		773.1	2011
279.4	249.2	1798.7	296.5	110.9		1325.5	2012
353.7		1647.4	408.4	125.9		1158.0	2013
404.2		1817.6	494.8	130.6		700.9	2014
512.6		1594.7	398.0	157.0		620.7	2015
489.9		1670.1	445.2	118.1	294.7	747.0	2016
450.8		1759.6	508.1	240.8	673.4	1391.0	2017
802.6		1854.7	470.5	298.5	619.2	1421.4	2018
803.7		1844.8	557.4	406.8	558.1	956.9	2019
1043.4		1626.3	862.6	705.8	453.6	1609.6	2020

Note: 1.For the year 2008 and before, there was data about investment in public transport.Starting from 2009, the data has only included investment in rail transport construction.

2.Starting from 2013, the national fixed assets Investment in the construction of municipal public utilities facilities has not included the fixed assets investment in urban flood prevention.

1-4-2 按行业分全国城市市政公用设施建设固定资产投资(2020年)

计量单位：万元

地区名称 Name of Regions	本年完成投资 Completed Investment of This Year	供 水 Water Supply	燃 气 Gas Supply	集中供热 Central Heating	轨道交通 Urban Rail Trainsit System	道路桥梁 Road and Bridge	地下综合管廊 Utility Tunnel
全 国 National Total	222839259	7494247	2386063	3938223	64208400	78143056	4535653
北 京 Beijing	15018987	270694	95627	239518	3635356	2892972	212376
天 津 Tianjin	4472897	126985	7360	30822	2889494	667000	1757
河 北 Hebei	5824553	211074	129400	351885	604245	1718421	618721
山 西 Shanxi	3442802	155717	42722	580106	100298	1548538	93554
内 蒙 古 Inner Mongolia	2218088	112517	2779	412670	458800	478401	52411
辽 宁 Liaoning	3100602	188441	91060	156943	980529	835338	
吉 林 Jilin	2494431	63683	67778	67660	574773	1110641	91956
黑 龙 江 Heilongjiang	2756836	352097	21593	194724	837374	743187	
上 海 Shanghai	4737785	277846	110168		1685400	1511068	17228
江 苏 Jiangsu	19750553	1084780	280326		6677754	6941313	158744
浙 江 Zhejiang	21536367	376283	112727		8921478	7874608	297598
安 徽 Anhui	8100502	302411	130817	30288	2152109	3005310	195598
福 建 Fujian	7772480	320497	90558		2832249	2243282	112312
江 西 Jiangxi	8267983	196158	64155		1636112	3107505	130602
山 东 Shandong	14813282	570940	177169	834392	2575469	6255528	220529
河 南 Henan	10283775	195274	91631	292446	3501312	2959581	165282
湖 北 Hubei	9905417	300569	46746	22699	3117944	3418080	700427
湖 南 Hunan	8110697	68837	144976		1716535	3560542	51379
广 东 Guangdong	16046482	494635	275941		4937442	5070595	323752
广 西 Guangxi	5602130	193880	84407		765689	2535208	364073
海 南 Hainan	796634	4322	179		24484	524460	14058
重 庆 Chongqing	9735406	362361	49010		2961139	4372267	60138
四 川 Sichuan	16261380	353994	64559		5190517	7307331	23795
贵 州 Guizhou	3433850	66248	9188	3000	1095700	1223693	22983
云 南 Yunnan	3567543	173833	25678		1228662	1008789	109913
西 藏 Tibet	123406	53				110312	
陕 西 Shaanxi	9342694	90515	63053	258842	2878471	3280428	321780
甘 肃 Gansu	1731720	125803	22701	84135	144598	658873	23223
青 海 Qinghai	290528	2791	6671	201		138901	49134
宁 夏 Ningxia	811741	190975	28384	159172		206974	17061
新 疆 Xinjiang	2024330	210247	39572	141776	84467	756576	60152
新疆生产建设兵团 Xinjiang Production and Construction Corps	463379	49785	9126	76945		77335	25116

National Fixed Assets Investment in Urban Service Facilities by Industry(2020)

Measurement Unit: 10,000RMB

排水 Sewerage	污水处理 Wastewater Treatment	污泥处置 Sludge Disposal	再生水利用 Wastewater Recycled and Reused	园林绿化 Landscaping	市容环境卫生 Environmental Sanitation	垃圾处理 Domestic Garbage Treatment	其他 Other	本年新增固定资产 Newly Added Fixed Assets of This Year	地区名称 Name of Regions
21147815	10130699	368561	303299	16262894	8626479	7057970	16096428	99164527	全　国
492303	118152		30744	2108501	205498	1314	4866142	3454315	北　京
259944	124307	767	11272	110128	337660	334519	41747	844938	天　津
744141	205358	34531	4190	767982	570850	496778	107834	2584971	河　北
133519	62478		400	440029	54152	44129	294167	1017249	山　西
232689	34684	1000	61901	141480	82370	55957	243970	644122	内　蒙　古
534108	120303	20113		72307	90588	77486	151288	1172531	辽　宁
260370	102889	1497	500	59740	46951	43304	150879	1190917	吉　林
327833	168128	16746	757	37865	148855	144240	93308	1158510	黑　龙　江
656790	607534	30000		135580	73381	73381	270324	1946176	上　海
1288994	656010	23571	21481	2030989	935347	846194	352306	8240125	江　苏
880725	592407	38965	1549	1444405	613912	537604	1014631	9930510	浙　江
1139228	402888	10116	26392	522066	364071	252492	258603	2418892	安　徽
1152463	784218	720	489	426251	362071	341093	232797	2425330	福　建
885777	311529	5109		564237	433840	277579	1249597	3365764	江　西
1714291	543810	33633	35448	1137375	395605	332055	931984	4797993	山　东
698594	269060	3260	5102	1717180	593653	482619	68821	7254403	河　南
1284270	295134	10094		426805	73513	65993	514363	3080003	湖　北
630945	507670	1091	282	310632	223679	210333	1403171	1679841	湖　南
3184359	2396940	16181	1535	248786	1021509	973041	489463	5756953	广　东
892819	91237	60256		293324	209277	199575	263453	3738715	广　西
62352	18069			26059	56861	24586	83858	131006	海　南
599239	87282	2835	568	837977	321519	306885	171756	5422203	重　庆
1411669	748266	12335	17827	958132	569378	408486	382005	18882766	四　川
143028	30291		7165	35870	89178	87955	744962	488740	贵　州
268812	112736	17023	3500	262110	138752	129350	350995	2676760	云　南
3041				10000				29160	西　藏
507291	218895	6920	345	751768	342274	114987	848272	1818558	陕　西
278793	242824	3000	13519	110130	37652	12178	245813	1191923	甘　肃
23888	10658		914	29723	10664		28555	148155	青　海
134053	77999		4995	37658	1610	502	35854	272206	宁　夏
244654	144992	17000	43118	182938	166888	148686	137060	1121529	新　疆
76834	43951	1797	9306	24865	54922	34671	68451	279262	新疆生产建设兵团

1-5-1　按资金来源分全国历年城市市政公用设施建设固定资产投资

计量单位：亿元

年 份 Year	本年资金 来源合计 Completed Investment of This Year	上 年 末 结余资金 The Balance of The Previous Year	本年资金来源		
			小 计 Subtotal	中央财政 拨 款 Financial Allocation From Central Government Budget	地方财政 拨 款 Financial Allocation From Local Government Budget
1978	12.0		6.4	6.4	
1979	14.0				
1980	14.4		14.4	6.1	
1981	20.0		20.0	5.3	
1982	27.2		27.2	8.6	
1983	28.2		28.2	8.2	
1984	41.7		41.7	11.8	
1985	63.8		63.8	13.9	
1986	79.8		79.8	13.2	
1987	90.0		90.0	13.4	
1988	112.6		112.6	10.2	
1989	106.8		106.8	9.7	
1990	121.2		121.2	7.4	
1991	169.9		169.9	8.6	
1992	265.4		265.4	9.9	
1993	521.6		521.6	15.9	
1994	665.5		665.5	27.9	
1995	837.0		807.5	24.2	
1996	939.1	68.0	871.1	34.8	
1997	1105.6	49.0	1056.6	43.0	
1998	1404.4	58.0	1346.4	100.2	
1999	1534.2	81.0	1453.2	173.8	
2000	1849.5	109.0	1740.5	222.0	
2001	2351.9	109.0	2112.8	104.9	379.1
2002	3123.2	111.0	2705.9	96.3	516.9
2003	4264.1	120.7	4143.4	118.9	733.4
2004	4650.9	267.9	4383.0	63.0	938.4
2005	5505.5	229.0	5276.6	63.9	1050.6
2006	5800.6	365.4	5435.2	89.2	1339.0
2007	6283.5	369.5	5914.0	77.3	1925.7
2008	7277.4	386.9	6890.4	72.7	2143.9
2009	10938.1	460.4	10477.6	112.9	2705.1
2010	13351.7	659.3	12692.4	206.0	3523.6
2011	14158.1	648.9	13509.1	166.3	4555.6
2012	15264.2	595.4	14668.9	171.1	4446.6
2013	16121.9	987.5	15134.3	147.5	3573.2
2014	16054.0	954.0	15100.0	102.2	4135.2
2015	16570.7	1295.0	15275.8	202.1	4406.4
2016	17319.2	942.7	16376.5	119.4	5183.7
2017	19704.7	1459.6	18245.1	290.4	5465.7
2018	19084.8	1245.5	17839.3	255.2	4552.3
2019	20438.8	1568.1	18870.7	412.6	5456.9
2020	23265.7	2228.6	21037.1	568.6	5922.0

注：自2013年起，"本年资金来源合计"为"本年实际到位资金合计"。

National Fixed Assets Investment in Urban Service Facilities by Capital Source in Past Years

Measurement Unit: 100 million RMB

| Sources of Fund | | | | | 年 份 |
国内贷款 Domestic Loan	债 券 Securities	利用外资 Foreign Investment	自筹资金 Self-Raised Funds	其他资金 Other Funds	 Year
					1978
					1979
0.1			8.2		1980
0.4			13.8	0.5	1981
1.0			16.5	1.1	1982
0.6			17.2	2.2	1983
2.1			25.5	2.3	1984
3.1		0.1	40.9	5.8	1985
3.1		0.1	57.4	6.0	1986
6.2		1.3	61.4	7.7	1987
7.1		1.6	78.0	15.7	1988
6.0		1.5	72.9	16.7	1989
11.0		2.2	82.2	18.4	1990
25.3		6.0	108.1	21.9	1991
42.9		10.3	180.4	38.9	1992
72.8		20.8	304.5	107.6	1993
58.3		64.2	397.1	118.0	1994
65.1		84.9	493.3	140.0	1995
122.2	4.9	105.6	486.1	117.5	1996
165.3	3.1	129.5	554.3	161.4	1997
284.8	40.3	110.1	600.4	210.6	1998
357.8	55.9	68.6	595.2	201.9	1999
428.6	29.0	76.7	682.7	301.5	2000
603.4	16.8	97.8	636.4	274.5	2001
743.8	7.3	109.6	866.3	365.7	2002
1435.4	17.4	90.0	1350.2	398.0	2003
1468.0	8.5	87.2	1372.9	445.0	2004
1805.9	5.2	170.0	1728.0	453.0	2005
1880.5	16.4	92.9	1638.1	379.2	2006
1763.7	29.5	73.1	1635.7	409.0	2007
2037.0	27.8	91.2	1980.1	537.6	2008
4034.8	120.8	66.1	2487.1	950.7	2009
4615.6	49.1	113.8	3058.9	1125.3	2010
3992.8	111.6	100.3	3478.6	1103.9	2011
4366.7	26.8	150.8	3740.5	1766.4	2012
4218.0	41.5	62.2	4714.1	2377.7	2013
4383.1	96.0	42.0	4294.7	2046.7	2014
3986.3	189.1	46.6	4258.0	2187.3	2015
4338.7	133.4	34.6	3963.6	2603.0	2016
4987.5	163.2	28.7	4997.8	2311.7	2017
4509.0	117.2	46.9	5105.2	3253.5	2018
5039.3	392.2	49.9	4707.3	2812.5	2019
3932.6	1864.0	67.5	5343.3	3339.1	2020

Note: Since 2013, Completed Investment of This Year is changed to be The Total Funds Actually Available for The Reported Year.

1-5-2　按资金来源分全国城市市政公用设施
建设固定资产投资(2020年)

计量单位: 万元

地区名称 Name of Regions	本年实际到位资金合计 The Total Funds Actually Available for The Reported Year	上年末结余资金 The Balance of The Previous Year	本年资金来源		
			小计 Subtotal	国家预算资金 State Budgetary Fund	国内贷款 Domestic Loan
				中央预算资金 Central Budgetary Fund	
全　国　**National Total**	**232656828**	**22285658**	**210371170**	**64905851**	**5685968**	**39325915**
北　京　Beijing	12731462	1792145	10939317	5405397	90027	2024633
天　津　Tianjin	4682331	382414	4299917	1335163	36017	1547833
河　北　Hebei	6692434	745311	5947123	2285682	346302	715156
山　西　Shanxi	3512163	166609	3345554	1446353	166535	468123
内　蒙　古　Inner Mongolia	2290045	95236	2194809	650191	88383	51590
辽　宁　Liaoning	3068439	132279	2936160	608844	224345	446298
吉　林　Jilin	4338837	782973	3555863	1290168	171336	351129
黑　龙　江　Heilongjiang	3383821	268703	3115118	883819	358894	528702
上　海　Shanghai	4879145	277550	4601595	2406527	19722	153951
江　苏　Jiangsu	23474665	2451363	21023302	4827356	102929	3199911
浙　江　Zhejiang	23095065	2881515	20213550	4958109	183744	4812201
安　徽　Anhui	7993336	453639	7539697	5006218	168907	273179
福　建　Fujian	7563005	613796	6949210	1728703	66703	1354051
江　西　Jiangxi	8847000	156654	8690346	3333041	223079	485077
山　东　Shandong	13030879	529494	12501385	3362558	27904	2244884
河　南　Henan	11889324	1671132	10218191	4297153	451427	2521887
湖　北　Hubei	9181678	993506	8188172	2713729	181213	2349745
湖　南　Hunan	7874360	1943569	5930791	887821	235340	410791
广　东　Guangdong	15406570	2134767	13271803	3376284	238766	3167085
广　西　Guangxi	6318533	397639	5920894	1525396	328844	1514341
海　南　Hainan	1044433	192664	851769	445941	11675	51346
重　庆　Chongqing	10431287	368785	10062502	3904897	163802	2834637
四　川　Sichuan	17040848	1320006	15720842	3169205	191139	3814788
贵　州　Guizhou	2544455	51909	2492546	112598	42138	262297
云　南　Yunnan	3361965	171570	3190395	429450	179728	929052
西　藏　Tibet	276254	2416	273838	16506	14147	
陕　西　Shaanxi	11127304	945031	10182273	2597684	495378	1982398
甘　肃　Gansu	2142742	201278	1941463	353522	163580	500705
青　海　Qinghai	565778	71946	493832	344728	123072	7467
宁　夏　Ningxia	886840	18667	868173	304452	62986	125844
新　疆　Xinjiang	2316673	22877	2293796	637266	292114	184561
新疆生产建设兵团　Xinjiang Production and Construction Corps	665159	48214	616945	261089	235793	12254

National Fixed Assets Investment in Urban Service Facilities by Capital Source(2020)

Measurement Unit: 10,000 RMB

债　券 Securities	利用外资 Foreign Investment	自筹资金 Self-Raised Funds	其他资金 Other Funds	各项应付款 Sum Payable This Year	地区名称 Name of Regions
18639941	675471	53433171	33390821	31652511	全　国
		1606555	1902732	1847861	北　京
481879	129595	618951	186496	224708	天　津
1501176		959096	486013	887366	河　北
209042		756300	465737	272193	山　西
305683	28673	873359	285313	414704	内 蒙 古
757650	17352	897714	208302	951913	辽　宁
898096	2951	811108	202411	315842	吉　林
1024190	31361	491502	155545	659605	黑 龙 江
		1955517	85600	998792	上　海
624374	4000	8574599	3793061	2665251	江　苏
502939	565	7024660	2915076	1849129	浙　江
75225	11323	1183080	990672	1777391	安　徽
1401606		1274592	1190258	512219	福　建
272227	170400	2547576	1882025	400048	江　西
1648982	61041	2593902	2590017	3655915	山　东
379996	75437	2389707	554012	1538755	河　南
373940	675	1721722	1028361	1385553	湖　北
441859	20000	2016492	2153828	196431	湖　南
1703826		1981903	3042704	1142485	广　东
178955	20273	1712301	969628	1139705	广　西
185888		29041	139553	161957	海　南
1824911	1491	1007058	489508	1862948	重　庆
1685864	39771	2350800	4660414	2452716	四　川
73347		1497992	546312	1030805	贵　州
304281	4000	1153649	369963	814578	云　南
44000		158333	55000		西　藏
161395	20	4431485	1009290	1040119	陕　西
261920	27838	185941	611536	505594	甘　肃
96641		16639	28357	25657	青　海
104676		259171	74030	408516	宁　夏
843922	28704	312073	287270	434652	新　疆
271451		40354	31797	79103	新疆生产建设兵团

二、居民生活数据
Data by Residents Living

1-6-1　全国历年城市供水情况

年 份 Year	综　合 生产能力 （万立方米／日） Integrated Production Capacity (10,000 m³/day)	供水管道 长　度 （公里） Length of Water Supply Pipelines (km)	供水总量 （万立方米） Total Quantity of Water Supply (10,000 m³)	生活用量 Residential Use
1978	2530.4	35984	787507	275854
1979	2714.0	39406	832201	309206
1980	2979.0	42859	883427	339130
1981	3258.0	46966	969943	367823
1982	3424.9	51513	1011319	391422
1983	3539.0	56852	1065956	421968
1984	3960.9	62892	1176474	465651
1985	4019.7	67350	1280238	519493
1986	10407.9	72557	2773921	706971
1987	11363.6	77864	2984697	759702
1988	12715.8	86231	3385847	873800
1989	12821.1	92281	3936648	930619
1990	14220.3	97183	3823425	1001021
1991	14584.0	102299	4085073	1159929
1992	16036.4	111780	4298437	1172919
1993	16927.9	123007	4502341	1282543
1994	18215.1	131052	4894620	1422453
1995	19250.4	138701	4815653	1581451
1996	19990.0	202613	4660652	1670673
1997	20565.8	215587	4767788	1757157
1998	20991.8	225361	4704732	1810355
1999	21551.9	238001	4675076	1896225
2000	21842.0	254561	4689838	1999960
2001	22900.0	289338	4661194	2036492
2002	23546.0	312605	4664574	2131919
2003	23967.1	333289	4752548	2246665
2004	24753.0	358410	4902755	2334625
2005	24719.8	379332	5020601	2437374
2006	26965.6	430426	5405246	2220459
2007	25708.4	447229	5019488	2263676
2008	26604.1	480084	5000762	2274266
2009	27046.8	510399	4967467	2334082
2010	27601.5	539778	5078745	2371488
2011	26668.7	573774	5134222	2476520
2012	27177.3	591872	5230326	2572473
2013	28373.4	646413	5373022	2676463
2014	28673.3	676727	5466613	2756911
2015	29678.3	710206	5604728	2872695
2016	30320.7	756623	5806911	3031376
2017	30475.0	797355	5937591	3153968
2018	31211.8	865017	6146244	3300567
2019	30897.8	920082	6283010	3401160
2020	32072.7	1006910	6295420	3484644

注：1.1978年至1985年综合供水生产能力为系统内数；1978年至1995年供水管道长度为系统内数。

　　2.自2006年起，供水普及率指标按城区人口和城区暂住人口合计为分母计算，括号中的数据为往年同口径数据。

National Urban Water Supply in Past Years

用水人口 （万人） Population with Access to Water Supply (10,000 persons)	人 均 日 生活用水量 （升） Daily Water Consumption Per Capita (liter)	供 水 普及率 (%) Water Coverage Rate (%)	年 份 Year
6267.1	120.6	81.6	1978
6951.0	121.8	82.3	1979
7278.0	127.6	81.4	1980
7729.3	130.4	53.7	1981
8102.2	132.4	56.7	1982
8370.9	138.1	52.5	1983
8900.7	143.3	49.5	1984
9424.3	151.0	45.1	1985
11757.9	161.9	51.3	1986
12684.6	164.1	50.4	1987
14049.9	170.4	47.6	1988
14786.3	172.4	47.4	1989
15611.1	175.7	48.0	1990
16213.2	196.0	54.8	1991
17280.8	186.0	56.2	1992
18636.4	188.6	55.2	1993
20083.0	194.0	56.0	1994
22165.7	195.4	58.7	1995
21997.0	208.1	60.7	1996
22550.1	213.5	61.2	1997
23169.1	214.1	61.9	1998
23885.7	217.5	63.5	1999
24809.2	220.2	63.9	2000
25832.8	216.0	72.26	2001
27419.9	213.0	77.85	2002
29124.5	210.9	86.15	2003
30339.7	210.8	88.85	2004
32723.4	204.1	91.09	2005
32304.1	188.3	86.67(97.04)	2006
34766.5	178.4	93.83	2007
35086.7	178.2	94.73	2008
36214.2	176.6	96.12	2009
38156.7	171.4	96.68	2010
39691.3	170.9	97.04	2011
41026.5	171.8	97.16	2012
42261.4	173.5	97.56	2013
43476.3	173.7	97.64	2014
45112.6	174.5	98.07	2015
46958.4	176.9	98.42	2016
48303.5	178.9	98.30	2017
50310.6	179.7	98.36	2018
51778.0	180.0	98.78	2019
53217.4	179.4	98.99	2020

Note: 1.Integrated production capacity from 1978 to 1985 is limited to the statistical figure in building sector; Length of water supply
pipelines from 1978 to 1995 is limited to the statistical figure in building sector.
2.Since 2006,water coverage rate has been calculated based on denominator which combines both permanent and temporary
residents in urban areas, and the data in brackets are the same index but calculated by the method of past years.

1-6-2 城市供水(2020年)

地区名称 Name of Regions	综合生产能力 (万立方米/日) Integrated Production Capacity (10,000m³/day)	地下水 Underground Water	供水管道长度 (公里) Length of Water Supply Pipelines (km)	建成区 in Built District	供水总量 (万立方米) Total Quantity of Water Supply (10,000m³)	生产运营用水 The Quantity of Water for Production and Operation
全 国 National Total	32072.65	5180.14	1006910.02	862347.85	6295419.56	1563871.70
北 京 Beijing	1788.16	1380.51	20535.10	11242.71	147792.65	15255.76
天 津 Tianjin	480.70	16.00	21613.19	20945.88	96013.05	29495.43
河 北 Hebei	831.63	341.74	22874.22	22204.68	157497.21	43264.25
山 西 Shanxi	420.87	224.91	13821.58	12371.38	87146.44	18830.66
内 蒙 古 Inner Mongolia	401.83	233.83	12732.70	11990.96	84486.16	29561.87
辽 宁 Liaoning	1222.24	290.64	39698.59	35842.97	266231.92	62926.69
吉 林 Jilin	655.63	77.88	16877.97	15589.91	105173.30	24572.74
黑 龙 江 Heilongjiang	667.39	224.14	22516.47	21892.61	137320.00	35679.40
上 海 Shanghai	1221.00		39552.54	39552.54	288576.67	41367.30
江 苏 Jiangsu	3490.41	60.72	119679.87	87760.44	587581.31	183944.80
浙 江 Zhejiang	2039.12	0.63	90784.64	64607.07	424103.51	140118.35
安 徽 Anhui	1079.04	94.96	36256.64	35348.28	238126.29	70810.93
福 建 Fujian	889.99	20.49	31722.18	31279.91	188912.54	38226.25
江 西 Jiangxi	646.33	1.80	27958.23	25104.12	143134.47	30025.15
山 东 Shandong	1908.07	584.42	59513.17	56415.71	379081.76	155140.30
河 南 Henan	1256.52	403.03	29138.19	27057.92	217730.42	56027.65
湖 北 Hubei	1597.28	5.24	51100.27	42296.07	302325.88	74638.35
湖 南 Hunan	1154.79	25.53	38420.44	29894.15	226436.32	40782.87
广 东 Guangdong	3949.10	26.91	128119.00	112096.36	956866.57	224641.91
广 西 Guangxi	708.04	38.57	23871.09	23140.95	185974.06	37330.96
海 南 Hainan	198.96	19.14	6984.19	990.43	50765.60	3025.00
重 庆 Chongqing	711.01	1.80	26524.11	24786.75	165351.83	35766.43
四 川 Sichuan	1669.27	69.59	52541.77	48131.36	291782.16	37678.90
贵 州 Guizhou	432.74	10.30	17303.53	9731.57	89270.51	17621.74
云 南 Yunnan	460.78	21.33	16734.77	13760.56	104904.32	18136.50
西 藏 Tibet	68.45	51.65	1771.42	1744.72	14700.23	1541.36
陕 西 Shaanxi	577.78	219.69	11801.88	11200.97	129087.24	35832.54
甘 肃 Gansu	373.80	84.45	6615.05	6200.12	56911.60	14857.40
青 海 Qinghai	140.80	120.81	3445.58	3043.74	30926.78	14552.21
宁 夏 Ningxia	277.22	104.07	3015.89	2944.83	35344.28	8135.26
新 疆 Xinjiang	547.91	269.07	11484.49	11284.92	89034.65	17666.59
新疆生产建设兵团 Xinjiang Production and Construction Corps	205.79	156.29	1901.26	1893.26	16829.83	6416.15

Urban Water Supply(2020)

公共服务 用　水 The Quantity of Water for Public Service	居民家庭 用　水 The Quantity of Water for Household Use	其他用水 The Quantity of Other Purposes	用水户数 （户） Number of Households with Access to Water Supply (unit)	家庭用户 Household User	用水人口 （万人） Population with Access to Water Supply (10,000 persons)	地区名称 Name of Regions
885907.42	2586607.85	306464.16	194395555	175503631	53217.37	全　　国
43131.67	62988.73	3970.63	6653840	6449944	1885.64	北　　京
13713.20	35256.15	3533.50	4997836	4824450	1174.44	天　　津
22426.95	67528.45	3239.32	6161814	5773822	1950.30	河　　北
15090.77	43819.43	1690.82	2593881	2443942	1207.60	山　　西
8128.52	25784.64	6390.08	4518534	3996349	917.62	内　蒙　古
35507.76	85094.67	17032.09	13312385	12461731	2251.61	辽　　宁
15409.19	36038.35	4512.64	6066807	5463058	1162.97	吉　　林
23130.50	42655.30	4897.48	7049628	6457917	1401.46	黑　龙　江
66982.41	113743.86	13786.98	9366490	8816530	2428.14	上　　海
74058.81	209636.35	43492.65	17348179	15882356	3538.02	江　　苏
60154.96	167139.70	11047.18	10873864	9761834	2833.02	浙　　江
28698.29	99120.71	7071.09	8070816	7379935	1775.18	安　　徽
29337.70	79223.00	13783.74	4902384	4336742	1388.12	福　　建
17474.07	65998.18	4875.27	5593672	5022232	1308.08	江　　西
43712.51	129449.63	11861.60	11576387	10656110	3982.77	山　　东
25013.93	98436.85	7968.03	7041015	6484614	2630.80	河　　南
27437.53	131879.32	7096.58	7760501	6804777	2273.46	湖　　北
31377.52	101553.26	7582.70	6028720	5312977	1738.39	湖　　南
147581.80	391233.23	60857.76	15054254	12873813	6241.10	广　　东
26854.84	94816.40	3096.59	3134876	2860987	1266.45	广　　西
6704.64	27916.87	5570.58	406554	347852	344.89	海　　南
24760.03	75286.53	5406.97	7354805	6738542	1524.60	重　　庆
43719.42	153405.41	11558.06	10015300	8648629	2760.15	四　　川
7394.09	43775.68	1722.78	3348219	2995954	828.17	贵　　州
7315.76	49550.14	10703.15	3571424	3104522	1007.82	云　　南
1450.16	9042.77	700.87	230023	185020	98.94	西　　藏
5877.17	66119.93	6457.48	3889360	3600613	1267.17	陕　　西
7545.07	24878.70	4892.09	1370953	1284739	636.43	甘　　肃
939.70	9786.27	3244.69	756865	653862	212.78	青　　海
5880.72	11711.81	4306.77	1059990	931741	296.50	宁　　夏
15857.01	30534.64	12337.33	3956878	2639641	780.42	新　　疆
3240.72	3202.89	1776.66	329301	308396	104.33	新疆生产 建设兵团

1-6-3 城市供水(公共供水)(2020年)

地区名称 Name of Regions	综合生产能力 (万立方米/日) Integrated Production Capacity (10,000 m³/day)	地下水 Under-ground Water	水厂个数 (个) Number of Water Plants (unit)	地下水 Under-ground Water	供水管道长度 (公里) Length of Water Supply Pipelines (km)	供水总量（万立方米）		
						合计 Total	售水量	
							小计 Subtotal	生产运营用水 The Quantity of Water for Production and Operation
全 国 **National Total**	**27625.01**	**2958.91**	**2943**	**848**	**980879.52**	**5864541.90**	**4911973.47**	**1238278.59**
北 京 Beijing	606.13	198.48	64	44	17819.39	134904.14	112458.28	10742.12
天 津 Tianjin	430.10	16.00	28	7	21598.19	90327.95	76313.18	26679.07
河 北 Hebei	730.82	267.69	144	87	20614.59	136725.01	115686.77	26179.58
山 西 Shanxi	365.29	178.58	80	62	12637.59	80058.92	72344.16	13910.99
内 蒙 古 Inner Mongolia	341.84	178.48	87	82	12457.28	71813.05	57192.00	20951.82
辽 宁 Liaoning	1057.40	216.64	156	74	37544.98	242591.92	176921.21	46697.71
吉 林 Jilin	371.36	41.22	68	23	15699.32	85825.26	61184.88	8717.66
黑 龙 江 Heilongjiang	573.01	154.13	102	62	21645.83	118332.80	87375.48	21352.74
上 海 Shanghai	1221.00		38		39552.54	288576.67	235880.55	41367.30
江 苏 Jiangsu	2879.50	31.30	130	14	118279.57	525640.41	449191.71	132381.02
浙 江 Zhejiang	1899.30		131		89740.41	405731.55	360088.23	122533.63
安 徽 Anhui	854.00	60.50	77	9	34837.27	203633.42	171208.15	43938.50
福 建 Fujian	884.02	18.60	84	9	31636.68	187797.88	159456.03	37478.86
江 西 Jiangxi	637.70	0.50	74	1	27846.14	142079.86	117318.06	29335.84
山 东 Shandong	1571.78	383.83	273	106	56515.66	311315.24	272397.52	100520.80
河 南 Henan	1046.79	235.50	136	62	27566.42	186782.60	156498.64	35871.63
湖 北 Hubei	1360.30		122		49233.02	281101.18	219827.08	56730.12
湖 南 Hunan	1019.90	21.50	96	8	38355.64	224664.82	179524.85	39643.50
广 东 Guangdong	3900.82	22.50	260	7	127966.42	952837.64	820285.77	220786.16
广 西 Guangxi	644.90	30.00	76	8	22653.29	172551.98	148676.71	24838.31
海 南 Hainan	190.74	14.70	17	1	6753.20	47894.76	40346.25	1485.87
重 庆 Chongqing	642.07		96		25844.50	162120.70	137988.83	33344.78
四 川 Sichuan	1616.41	50.70	163	22	52371.52	287301.29	241880.92	34570.82
贵 州 Guizhou	432.74	10.30	86	3	17303.53	89270.51	70514.29	17621.74
云 南 Yunnan	443.66	13.15	114	12	16677.36	99986.47	80787.70	15319.91
西 藏 Tibet	63.22	51.42	16	11	1771.42	14524.19	12559.12	1541.32
陕 西 Shaanxi	463.36	147.44	80	47	10736.91	112053.17	97253.05	25871.16
甘 肃 Gansu	341.72	79.41	42	24	6301.95	54792.74	50054.40	13793.78
青 海 Qinghai	82.31	74.21	13	9	3056.71	20055.70	17651.79	4840.82
宁 夏 Ningxia	232.90	70.55	23	16	2633.89	30745.64	25435.92	6005.99
新 疆 Xinjiang	514.13	235.29	55	33	11327.04	85674.60	73035.52	16808.89
新疆生产建设兵团 Xinjiang Production and Construction Corps	205.79	156.29	12	5	1901.26	16829.83	14636.42	6416.15

Urban Water Supply(Public Water Suppliers)(2020)

Total Quantity of Water Supply(10,000m³)						用水户数 （户） Number of Households with Access to Water Supply (unit)	家庭用户 Household User	用水人口 （万人） Population with Access to Water Supply (10,000 persons)	地区名称 Name of Regions
Water Sold			免费 供水量	生活用水	漏损水量				
公共服务 用　水 The Quantity of Water for Public Service	居民家庭 用　水 The Quantity of Water for Household Use	其他用水 The Quantity of Other Purposes	The Quantity of Free Water Supply	Domestic Water Use	The Lossed Water				
841289.08	2546264.82	286140.98	167160.32	12128.52	785408.11	191767415	173698354	52599.83	全　　国
36960.50	62197.72	2557.94	695.41	5.00	21750.45	6507150	6314197	1768.00	北　　京
13298.92	32801.69	3533.50	1849.99	623.06	12164.78	4996234	4824448	1174.43	天　　津
21543.01	65136.55	2827.63	3353.52	695.16	17684.72	6024022	5653081	1923.18	河　　北
14667.69	42339.61	1425.87	467.39	91.27	7247.37	2285740	2152046	1175.87	山　　西
5436.93	24573.23	6230.02	3266.81		11354.24	4468171	3950786	897.54	内　蒙　古
31586.18	83432.78	15204.54	22820.48	2653.17	42850.23	13103923	12352141	2222.28	辽　　宁
13999.31	34771.09	3696.82	4449.23	251.79	20191.15	5984328	5401628	1140.86	吉　　林
21032.12	41289.52	3701.10	2917.11	444.54	28040.21	6935756	6349528	1380.10	黑　龙　江
66982.41	113743.86	13786.98	8242.65		44453.47	9366490	8816530	2428.14	上　　海
68940.58	208261.56	39608.55	9659.23	1293.62	66789.47	17288650	15842450	3525.34	江　　苏
59945.17	166562.25	11047.18	4807.48	260.44	40835.84	10831143	9721102	2820.37	浙　　江
24218.50	96127.76	6923.39	5667.69	67.90	26757.58	8007941	7322719	1763.70	安　　徽
29309.65	78883.78	13783.74	6240.24	101.49	22101.61	4897261	4335642	1387.87	福　　建
17415.08	65809.93	4757.21	3486.91	882.59	21274.89	5584251	5014443	1306.28	江　　西
38786.33	123728.00	9362.39	3449.53	345.25	35468.19	11245657	10396010	3897.01	山　　东
20166.99	94761.11	5698.91	5384.25	413.34	24899.71	6861728	6336786	2559.04	河　　南
26620.66	129810.34	6665.96	16041.72	712.32	45232.38	7313286	6730552	2255.22	湖　　北
31098.57	101238.76	7544.02	11543.87	1250.08	33596.10	6020853	5308993	1733.51	湖　　南
147579.74	391183.85	60736.02	18289.72	573.63	114262.15	15040589	12862303	6235.05	广　　东
26482.90	94346.32	3009.18	2743.35	125.78	21131.92	3123788	2851374	1260.06	广　　西
6590.68	27133.00	5136.70	1844.91	30.55	5703.60	392275	347151	340.48	海　　南
24729.74	74633.25	5281.06	3875.30	8.00	20256.57	7330555	6716941	1507.43	重　　庆
42511.68	153307.80	11490.62	5602.08	730.02	39818.29	9992487	8630407	2754.82	四　　川
7394.09	43775.68	1722.78	6063.14	152.81	12693.08	3348219	2995954	828.17	贵　　州
7211.76	47556.89	10699.14	6168.24	200.47	13030.53	3562801	3096377	1000.93	云　　南
1390.16	8962.77	664.87	493.83	10.99	1471.24	229820	184817	98.84	西　　藏
4594.09	62225.81	4561.99	2759.58	19.67	12040.54	3679043	3496945	1206.40	陕　　西
7256.18	24337.31	4667.13	817.03	93.56	3921.31	1321782	1236005	623.25	甘　　肃
939.70	8646.58	3224.69	389.14	6.65	2014.77	738988	636513	209.92	青　　海
4501.46	11257.91	3670.56	1764.60	0.80	3545.12	1023317	896622	293.10	宁　　夏
14857.58	30225.22	11143.83	1698.79	64.02	10940.29	3931866	2615467	778.31	新　　疆
3240.72	3202.89	1776.66	307.10	20.55	1886.31	329301	308396	104.33	新疆生产 建设兵团

1-6-4　城市供水(自建设施供水)(2020年)

地区名称 Name of Regions	综　合 生产能力 (万立方米/日) Integrated Production Capacity (10,000 m³/day)	地下水 Underground Water	供水管道 长　度 (公里) Length of Water Supply Pipelines (km)	建成区 in Built District	供水总量(万立方米)	
					合计 Total	生产运营 用　水 The Quantity of Water for Production and Operation
全　国　National Total	**4447.64**	**2221.23**	**26030.50**	**16958.78**	**430877.66**	**325593.11**
北　京　Beijing	1182.03	1182.03	2715.71		12888.51	4513.64
天　津　Tianjin	50.60		15.00	15.00	5685.10	2816.36
河　北　Hebei	100.81	74.05	2259.63	1942.48	20772.20	17084.67
山　西　Shanxi	55.58	46.33	1183.99	455.67	7087.52	4919.67
内　蒙　古　Inner Mongolia	59.99	55.35	275.42	173.42	12673.11	8610.05
辽　宁　Liaoning	164.84	74.00	2153.61	1503.08	23640.00	16228.98
吉　林　Jilin	284.27	36.66	1178.65	818.24	19348.04	15855.08
黑　龙　江　Heilongjiang	94.38	70.01	870.64	543.14	18987.20	14326.66
上　海　Shanghai						
江　苏　Jiangsu	610.91	29.42	1400.30	1242.88	61940.90	51563.78
浙　江　Zhejiang	139.82	0.63	1044.23	843.63	18371.96	17584.72
安　徽　Anhui	225.04	34.46	1419.37	1308.45	34492.87	26872.43
福　建　Fujian	5.97	1.89	85.50	85.50	1114.66	747.39
江　西　Jiangxi	8.63	1.30	112.09	91.22	1054.61	689.31
山　东　Shandong	336.29	200.59	2997.51	2335.75	67766.52	54619.50
河　南　Henan	209.73	167.53	1571.77	1023.10	30947.82	20156.02
湖　北　Hubei	236.98	5.24	1867.25	618.62	21224.70	17908.23
湖　南　Hunan	134.89	4.03	64.80	30.80	1771.50	1139.37
广　东　Guangdong	48.28	4.41	152.58	148.58	4028.93	3855.75
广　西　Guangxi	63.14	8.57	1217.80	1187.40	13422.08	12492.65
海　南　Hainan	8.22	4.44	230.99		2870.84	1539.13
重　庆　Chongqing	68.94	1.80	679.61	671.91	3231.13	2421.65
四　川　Sichuan	52.86	18.89	170.25	44.76	4480.87	3108.08
贵　州　Guizhou						
云　南　Yunnan	17.12	8.18	57.41	32.51	4917.85	2816.59
西　藏　Tibet	5.23	0.23			176.04	0.04
陕　西　Shaanxi	114.42	72.25	1064.97	935.82	17034.07	9961.38
甘　肃　Gansu	32.08	5.04	313.10	276.10	2118.86	1063.62
青　海　Qinghai	58.49	46.60	388.87	147.27	10871.08	9711.39
宁　夏　Ningxia	44.32	33.52	382.00	326.00	4598.64	2129.27
新　疆　Xinjiang	33.78	33.78	157.45	157.45	3360.05	857.70
新疆生产 建设兵团　Xinjiang Production and Construction Corps						

Urban Water Supply (Suppliers with Self-Built Facilities)(2020)

Total Quantity of Water Supply(10,000 m³)			用水户数 （户） Number of Households with Access to Water Supply (unit)	家庭用户 Household User	用水人口 （万人） Population with Access to Water Supply (10,000 persons)	地区名称 Name of Regions
公共服务 用　水 The Quantity of Water for Public Service	居民家庭 用　水 The Quantity of Water for Household Use	其他用水 The Quantity of Water for Other Purposes				
44618.34	**40343.03**	**20323.18**	**2628140**	**1805277**	**617.54**	全　　国
6171.17	791.01	1412.69	146690	135747	117.64	北　　京
414.28	2454.46		1602	2	0.01	天　　津
883.94	2391.90	411.69	137792	120741	27.12	河　　北
423.08	1479.82	264.95	308141	291896	31.73	山　　西
2691.59	1211.41	160.06	50363	45563	20.08	内　蒙　古
3921.58	1661.89	1827.55	208462	109590	29.33	辽　　宁
1409.88	1267.26	815.82	82479	61430	22.11	吉　　林
2098.38	1365.78	1196.38	113872	108389	21.36	黑　龙　江
						上　　海
5118.23	1374.79	3884.10	59529	39906	12.68	江　　苏
209.79	577.45		42721	40732	12.65	浙　　江
4479.79	2992.95	147.70	62875	57216	11.48	安　　徽
28.05	339.22		5123	1100	0.25	福　　建
58.99	188.25	118.06	9421	7789	1.80	江　　西
4926.18	5721.63	2499.21	330730	260100	85.76	山　　东
4846.94	3675.74	2269.12	179287	147828	71.76	河　　南
816.87	2068.98	430.62	447215	74225	18.24	湖　　北
278.95	314.50	38.68	7867	3984	4.88	湖　　南
2.06	49.38	121.74	13665	11510	6.05	广　　东
371.94	470.08	87.41	11088	9613	6.39	广　　西
113.96	783.87	433.88	14279	701	4.41	海　　南
30.29	653.28	125.91	24250	21601	17.17	重　　庆
1207.74	97.61	67.44	22813	18222	5.33	四　　川
						贵　　州
104.00	1993.25	4.01	8623	8145	6.89	云　　南
60.00	80.00	36.00	203	203	0.10	西　　藏
1283.08	3894.12	1895.49	210317	103668	60.77	陕　　西
288.89	541.39	224.96	49171	48734	13.18	甘　　肃
	1139.69	20.00	17877	17349	2.86	青　　海
1379.26	453.90	636.21	36673	35119	3.40	宁　　夏
999.43	309.42	1193.50	25012	24174	2.11	新　　疆
						新疆生产 建设兵团

1-7-1 全国历年城市节约用水情况

计量单位：万立方米

年 份 Year	计划用水量 Planned Quantity of Water Use	新水取用量 Fresh Water Used
1991	1977024	1921307
1992	2077092	1939390
1993	2076928	2004922
1994	2299596	2497236
1995	2123215	1930610
1996	6285246	2166806
1997	2004412	1875974
1998	2490390	2343370
1999	2409254	2223035
2000	2153979	2071731
2001	2504134	2126401
2002	2463717	2091365
2003	2353286	2014528
2004	2441914	2048488
2005	2676425	2300332
2006		2356268
2007		2295192
2008		2121034
2009		2064014
2010		2161262
2011		1694727
2012		1838306
2013		1876372
2014		1950107
2015		1922051
2016		1971043
2017		2455307
2018		2290261
2019		2333077
2020		2350149

注：自2006年起，不统计计划用水量指标。

National Urban Water Conservation in Past Years

Measurement Unit: 10,000 m³

工业用水 重复利用量 Quantity of Industrial Water Recycled	节约用水量 Water Saved	年 份 Year
1985326	211259	1991
2843918	205087	1992
3000109	218317	1993
3375652	276241	1994
3626086	235113	1995
3345351	235194	1996
4429175	260885	1997
3978967	278835	1998
3966355	287284	1999
3811945	353569	2000
3881683	377733	2001
4849164	372352	2002
4572711	338758	2003
4183937	393426	2004
5670096	376093	2005
5535492	415755	2006
6026048	454794	2007
6547258	659114	2008
6130645	628692	2009
6872928	407152	2010
6334101	406578	2011
7388185	400806	2012
6526517	382760	2013
6930588	405234	2014
7160130	403133	2015
7644277	576220	2016
8211738	648227	2017
8556264	508437	2018
9081103	499888	2019
10305444	707572	2020

Note: From 2006, "Planned Quantity of Water Use" has not been counted.

1-7-2　城市节约用水（2020年）

计量单位：万立方米

地区名称 Name of Regions	计划用水户数 （户） Planned Water Consumers (unit)	自备水计划用水户数 Planned Self-produced Water Consumers	计划用水户实际用水量		新水取水量		重复利用量
			合　计 Total	工　业 Industry	新水取水量 Fresh Water Used	工　业 Industry	重复利用量 Water Reused
全　国 National Total	**2281152**	**145063**	**13242300**	**11348304**	**2350149**	**1042860**	**10892151**
北　京 Beijing	40405	4940	209187	17442	209187	17442	
天　津 Tianjin	13410	1446	1177898	1161974	40658	26163	1137240
河　北 Hebei	42134	746	219790	205022	28036	14240	191754
山　西 Shanxi	7686	630	492006	437418	76069	28113	415937
内 蒙 古 Inner Mongolia	390894	369	224939	200333	50595	29121	174344
辽　宁 Liaoning	54662	179	768559	722768	55956	34149	712603
吉　林 Jilin	4527	586	163678	156560	34453	27611	129226
黑 龙 江 Heilongjiang	1904	1135	214348	189998	95672	72775	118677
上　海 Shanghai	140255		98027	37176	98027	37176	
江　苏 Jiangsu	26036	984	2053523	1614546	311835	177382	1741688
浙　江 Zhejiang	38684	1663	793541	723454	150947	108533	642594
安　徽 Anhui	6758	266	726832	693702	78810	47936	648021
福　建 Fujian	97880	5	136905	93498	56905	14465	80000
江　西 Jiangxi	1960	187	114769	95495	26754	7697	88015
山　东 Shandong	183846	66527	1088830	917018	179465	83965	909365
河　南 Henan	55283	11608	1074322	1036706	66188	35153	1008133
湖　北 Hubei	153483	87	656212	595747	92322	48731	563890
湖　南 Hunan	81388	12	183832	135299	75219	29559	108613
广　东 Guangdong	434260	6178	1188665	1085831	182869	80205	1005796
广　西 Guangxi	8966	176	446716	274158	93797	17232	352919
海　南 Hainan	15419	281	98506	16811	87322	5720	11184
重　庆 Chongqing	949	23	8517	5889	3533	912	4984
四　川 Sichuan	32621	4328	169165	115667	84163	35545	85002
贵　州 Guizhou	5545	136	66603	38726	31086	5980	35517
云　南 Yunnan	142326	546	114517	100894	24730	11530	89787
西　藏 Tibet	322		1200		1200		
陕　西 Shaanxi	44724	12426	53970	25565	40182	12051	13788
甘　肃 Gansu	242716	27001	504296	480475	46976	24955	457320
青　海 Qinghai	5	5	2387	987	865	412	1522
宁　夏 Ningxia	759	245	168367	162072	13088	6793	155279
新　疆 Xinjiang	6530	2348	13620	1800	12647	1080	973
新疆生产建设兵团 Xinjiang Production and Construction Corps	4815		8573	5274	595	235	7978

Urban Water Conservation (2020)

Measurement Unit: 10,000m^3

| Actual Quantity of Water Used | | | | 节约用水量 | | 节水措施投资总额（万元）Total Investment in Water-Saving Measures (10,000 RMB) | 地区名称 |
工　业 Industry	超计划定额用水量 Water Quantity Consumed in Excess of Quota	重复利用率 (%) Reuse Rate (%)	工　业 Industry	Water Saved	工　业 Industry		Name of Regions
10305444	20383	82.25	90.81	707572	470678	551148	全　　国
	89			10208	3707	6258	北　　京
1135811	184	96.55	97.75	67	37	2362	天　　津
190782	95	87.24	93.05	6190	4794	2888	河　　北
409305	641	84.54	93.57	15146	10720	3150	山　　西
171212	47	77.51	85.46	7373	4570	168	内　蒙　古
688619	115	92.72	95.28	12367	8550	16632	辽　　宁
128949	90	78.95	82.36	5924	5188	224	吉　　林
117223	52	55.37	61.70	16188	9808	2246	黑　龙　江
				81682	7843	12986	上　　海
1437164	2199	84.81	89.01	69704	54371	56150	江　　苏
614920	872	80.98	85.00	39065	29075	57970	浙　　江
645765	94	89.16	93.09	28204	23614	73593	安　　徽
79033	3774	58.43	84.53	19349	14721	6687	福　　建
87798	4	76.69	91.94	5469	5381	8617	江　　西
833053	1164	83.52	90.84	37324	26833	56341	山　　东
1001554	615	93.84	96.61	21149	13034	10838	河　　南
547016	745	85.93	91.82	36864	32138	31690	湖　　北
105740	723	59.08	78.15	100325	96716	16822	湖　　南
1005627	2465	84.62	92.61	109340	73210	113497	广　　东
256926	4373	79.00	93.71	4154	1589	2488	广　　西
11091	216	11.35	65.98	2202	170	11030	海　　南
4977		58.52	84.52	15	8	1956	重　　庆
80122	348	50.25	69.27	26014	18566	4028	四　　川
32746	428	53.33	84.56	19380	18999	5056	贵　　州
89364	682	78.40	88.57	7604	326	10261	云　　南
							西　　藏
13514	56	25.55	52.86	11904	3061	24354	陕　　西
455520	40	90.68	94.81	6105	1656	4006	甘　　肃
575	74	63.76	58.26	298	298	96	青　　海
155279	200	92.23	95.81	3065	1013	6745	宁　　夏
720		7.15	40.02	4892	679	2010	新　　疆
5039		93.06	95.55				新疆生产建设兵团

1-8-1 全国历年城市燃气情况
National Urban Gas in Past Years

年 份 Year	人工煤气 Man-Made Coal Gas				天然气 Natural Gas			
	供气总量 (万立方米) Total Gas Supplied (10,000 m³)	家庭用量 Domestic Consumption	用气人口 (万人) Population with Access to Gas (10,000 persons)	管道长度 (公里) Length of Gas Supply Pipeline (km)	供气总量 (万立方米) Total Gas Supplied (10,000 m³)	家庭用量 Domestic Consumption	用气人口 (万人) Population with Access to Gas (10,000 persons)	管道长度 (公里) Length of Gas Supply Pipeline (km)
1978	172541	66593	450	4157	69078	4103	24	560
1979	182748	73557	500	4446	68489	9833	76	751
1980	195491	83281	561	4698	58937	4893	80	921
1981	199466	90326	594	4830	93970	8575	83	1059
1982	208819	94896	630	5258	90421	8938	93	1098
1983	214009	89012	703	5967	49071	9443	91	1149
1984	231351	96228	768	7353	168568	38258	135	1965
1985	244754	107060	911	8255	162099	43268	281	2312
1986	337745	139374	951	7990	435887	65262	492	2409
1987	686518	170887	1177	11650	501093	77719	634	5465
1988	667927	171376	1357	13028	573983	73305	748	6186
1989	841297	204709	1498	14448	591169	91770	896	6849
1990	1747065	274127	1674	16312	642289	115662	972	7316
1991	1258088	311430	1867	18181	754616	154302	1072	8054
1992	1495531	305325	2181	20931	628914	152013	1119	8487
1993	1304130	345180	2490	23952	637207	139297	1180	8889
1994	1262712	416453	2889	27716	752436	146938	1273	9566
1995	1266894	456585	3253	33890	673354	163788	1349	10110
1996	1348076	472904	3490	38486	637832	138018	1470	18752
1997	1268944	535412	3735	41475	663001	177121	1656	22203
1998	1675571	480734	3746	42725	688255	195778	1908	25429
1999	1320925	494001	3918	45856	800556	215006	2225	29510
2000	1523615	630937	3944	48384	821476	247580	2581	33655
2001	1369144	494191	4349	50114	995197	247543	3127	39556
2002	1989196	490258	4541	53383	1259334	350479	3686	47652
2003	2020883	583884	4792	57017	1416415	374986	4320	57845
2004	2137225	512026	4654	56419	1693364	454248	5628	71411
2005	2558343	458538	4369	51404	2104951	521389	7104	92043
2006	2964500	381518	4067	50524	2447742	573441	8319	121498
2007	3223512	373522	4022	48630	3086365	662198	10190	155271
2008	3558287	353162	3370	45172	3680393	779917	12167	184084
2009	3615507	307134	2971	40447	4050996	913386	14544	218778
2010	2799380	268764	2802	38877	4875808	1171596	17021	256429
2011	847256	238876	2676	37100	6787997	1301190	19028	298972
2012	769686	215069	2442	33538	7950377	1558311	21208	342752
2013	627989	167886	1943	30467	8882417	1726620	23783	388466
2014	559513	145773	1757	29043	9643783	1968878	25973	434571
2015	471378	108306	1322	21292	10407906	2080061	28561	498087
2016	440944	108716	1085	18513	11717186	2864124	30856	551031
2017	270882	73733	752	11716	12637546	2825027	33934	623253
2018	297893	78957	779	13124	14439538	3135097	36902	698043
2019	276841	56168	675	10915	15279409	3470004	39025	767946
2020	231447	52031	548	9860	15637020	3815984	41302	850552

注：自2006年起，燃气普及率指标按城区人口和城区暂住人口合计为分母计算，括号中的数据为与往年同口径数据。

Note: Since 2006, gas coverage rate has been calculated based on denominator which combines both permanent and temporary residents in urban areas, and the data in brackets are the same index but calculated by the method of past years.

1-8-1 续表 continued

年 份 Year	液化石油气　LPG		用气人口 （万人） Population with Access to Gas (10,000 persons)	管道长度 （公里） Length of Gas Supply Pipeline (km)	燃气普及率 （%） Gas Coverage Rate (%)
	供气总量 （吨） Total Gas Supplied (ton)	家庭用量 Domestic Consumption			
1978	194533	175744	635		14.4
1979	243576	218797	788		16.1
1980	290460	269502	924		17.3
1981	330987	308028	995		11.6
1982	388596	342828	1076		12.6
1983	456192	414635	1159		12.3
1984	535289	424318	1435		13.0
1985	601803	540761	1534		13.0
1986	1011308	763590	2041		15.2
1987	1049116	954534	2399		16.7
1988	1730261	1136883	2764		16.5
1989	1898003	1259349	3156		17.8
1990	2190334	1428058	3579		19.1
1991	2423988	1694399	4084		23.7
1992	2996699	2019620	4796		26.3
1993	3150296	2316129	5770		27.9
1994	3664948	2817702	6745		30.4
1995	4886528	3701504	8355		34.3
1996	5758374	3943604	8864	2762	38.2
1997	5786023	4370979	9350	4086	40.0
1998	7972947	5478535	9995	4458	41.8
1999	7612684	4990363	10336	6116	43.8
2000	10537147	5322828	11107	7419	45.4
2001	9818313	5583497	13875	10809	60.42
2002	11363884	6561738	15431	12788	67.17
2003	11263475	7817094	16834	15349	76.74
2004	11267120	7041351	17559	20119	81.53
2005	12220141	7065214	18013	18662	82.08
2006	12636613	6936513	17100	17469	79.11(88.58)
2007	14667692	7280415	18172	17202	87.40
2008	13291072	6292713	17632	28590	89.55
2009	13400303	6887600	16924	14236	91.41
2010	12680054	6338523	16503	13374	92.04
2011	11658326	6329164	16094	12893	92.41
2012	11148032	6081312	15683	12651	93.15
2013	11097298	6130639	15102	13437	94.25
2014	10828490	5862125	14378	10986	94.57
2015	10392169	5871062	13955	9009	95.30
2016	10788042	5739456	13744	8716	95.75
2017	9988088	5447739	12616	6200	96.26
2018	10153298	5447936	11782	4841	96.70
2019	9227179	4917008	11297	4452	97.29
2020	8337109	4786679	10767	4010	97.87

年 份 Year	液化石油气　LPG		用气人口 （万人） Population with Access to Gas (10,000 persons)	管道长度 （公里） Length of Gas Supply Pipeline (km)	燃气普及率 （%） Gas Coverage Rate (%)
	供气总量 （吨） Total Gas Supplied (ton)	家庭用量 Domestic Consumption			

1-8-2　城市人工煤气(2020年)

地区名称 Name of Regions	生产能力 (万立方米／日) Production Capacity (10,000 m³)	储气能力 (万立方米) Gas Storage Capacity (10,000 m³)	供气管道 长　度 (公里) Length of Gas Supply Pipeline (km)	自制气量 (万立方米) Self-Produced Gas (10,000 m³)	合　计 Total
全　国　National Total	1394.61	259.66	9859.56	280212.62	231447.26
河　北　Hebei	11.79	11.50	813.67	4300.00	46895.83
山　西　Shanxi	262.00	27.00	989.90	33569.00	45340.21
内 蒙 古　Inner Mongolia			291.00		3170.87
辽　宁　Liaoning	178.12	95.00	3817.64	19853.92	31917.36
吉　林　Jilin		8.60	346.50		3158.00
黑 龙 江　Heilongjiang		13.50	318.90		2380.00
福　建　Fujian	10.00	5.00	156.00	883.60	883.60
江　西　Jiangxi	12.00	21.60	580.78	2018.00	14836.93
山　东　Shandong	340.00	1.00	10.00	60624.00	29487.00
河　南　Henan	179.50	30.80	241.56	31397.60	30581.60
湖　北　Hubei			589.11		
广　西　Guangxi	9.60	18.20	667.17		4626.99
四　川　Sichuan		18.40	700.83		14320.22
甘　肃　Gansu	345.60	7.00	276.00	125814.50	2092.10
新　疆　Xinjiang	46.00	2.00	60.50	1752.00	1752.50
新疆生产 建设兵团　Xinjiang Production and Construction Corps		0.06			4.05

Urban Man-Made Coal Gas(2020)

供气总量(万立方米) Total Gas Supplied (10,000 m³)		燃气损失量	用气户数 (户) Number of Households with Access to Gas (unit)	用气户数 家庭用户 Household User	用气人口 (万人) Population with Access to Gas (10,000 persons)	地区名称 Name of Regions
销售气量 Quantity Sold	居民家庭 Households	Loss Amount				
224382.15	52030.78	7065.11	2455014	2418430	548.20	全　　国
45671.88	2134.78	1223.95	127312	125643	28.26	河　　北
44469.47	2018.18	870.74	94686	94368	27.30	山　　西
3113.34	2445.78	57.53	65000	63000	14.43	内 蒙 古
29069.37	25218.95	2847.99	1439595	1413434	288.87	辽　　宁
3126.00	2031.00	32.00	165000	164085	41.25	吉　　林
2005.00	1752.00	375.00	107800	105645	33.85	黑 龙 江
868.60	868.60	15.00	19650	19650	6.80	福　　建
14030.17	1185.67	806.76	37890	36595	6.60	江　　西
29486.00		1.00	2			山　　东
30554.60	24.20	27.00	251	108	0.04	河　　南
						湖　　北
4488.18	3388.39	138.81	125916	125272	34.51	广　　西
13794.13	7397.50	526.09	189257	188137	45.32	四　　川
1949.36	1809.68	142.74	67152	66990	14.27	甘　　肃
1752.00	1752.00	0.50	15000	15000	4.50	新　　疆
4.05	4.05		503	503	2.20	新疆生产 建设兵团

1-8-3　城市天然气（2020年）

地区名称 Name of Regions	储气能力 （万立方米） Gas Storage Capacity (10,000 m³)	供气管道长度 （公里） Length of Gas Supply Pipeline (km)	供气总量(万立方米)			
			合　计 Total	销售气量 Quantity Sold	居民家庭 Households	集中供热 Central Heating
全　国　National Total	133031.16	850552.13	15637019.69	15302771.08	3815983.88	1559150.47
北　京　Beijing	594.78	30303.48	1854130.00	1779490.00	158446.00	561110.00
天　津　Tianjin	1149.68	47064.15	601575.74	589362.65	93414.14	159628.08
河　北　Hebei	5958.67	39547.28	642810.08	623401.27	202956.55	66288.45
山　西　Shanxi	389.31	24701.82	280869.23	273225.27	86482.98	22297.24
内　蒙　古　Inner Mongolia	6679.27	11064.64	207513.88	206151.70	83028.14	31454.67
辽　宁　Liaoning	2372.98	29121.88	339622.38	331054.81	72772.98	8834.74
吉　林　Jilin	928.54	13892.25	214185.86	209319.78	41805.00	3893.36
黑　龙　江　Heilongjiang	482.51	11560.09	153325.78	150615.41	40245.20	6215.54
上　海　Shanghai	72900.00	32864.95	899018.66	867347.96	173824.55	
江　苏　Jiangsu	7535.82	95314.80	1437793.73	1414415.74	310762.66	5993.21
浙　江　Zhejiang	2742.23	50862.51	775960.11	768710.52	124016.28	
安　徽　Anhui	3704.19	32274.37	414416.95	404615.07	128569.82	
福　建　Fujian	573.70	14737.76	258571.21	257369.34	29331.75	
江　西　Jiangxi	930.57	18243.87	194416.86	191495.10	66119.29	145.00
山　东　Shandong	3056.01	74784.30	1178044.54	1158075.90	267268.81	104321.28
河　南　Henan	1032.13	29996.85	620406.20	603849.19	225625.81	46273.61
湖　北　Hubei	4208.40	46215.20	548802.03	533983.24	166272.53	389.00
湖　南　Hunan	2041.23	24023.19	283930.05	276835.39	119480.08	16.59
广　东　Guangdong	2428.35	43629.13	1269926.89	1265063.83	165886.98	
广　西　Guangxi	915.88	9655.01	156756.88	156138.12	44230.72	
海　南　Hainan	214.05	4660.21	38796.31	38202.86	20797.06	
重　庆　Chongqing	432.27	24556.76	526163.95	511762.56	230166.31	2438.31
四　川　Sichuan	415.94	62616.52	865231.27	844355.84	390412.65	387.02
贵　州　Guizhou	1178.96	8860.84	126336.81	125233.67	42938.44	336.39
云　南　Yunnan	378.72	8708.45	60286.66	59730.38	18179.53	
西　藏　Tibet	100.00	6165.85	4268.78	4055.33	2217.92	1837.41
陕　西　Shaanxi	3621.13	23363.66	556358.80	545773.98	227766.13	142918.62
甘　肃　Gansu	313.74	4265.21	254404.96	253427.01	64666.40	75541.15
青　海　Qinghai	167.42	3295.83	170972.65	164766.13	47705.60	65273.76
宁　夏　Ningxia	67.50	7217.20	161213.93	159564.36	44841.01	21380.28
新　疆　Xinjiang	5296.98	15718.11	500066.93	495401.10	115534.78	228138.27
新疆生产建设兵团　Xinjiang Production and Construction Corps	220.20	1265.96	40841.58	39977.49	10217.78	4038.49

Urban Natural Gas(2020)

Total Gas Supplied (10,000 m³)		用气户数	家庭用户	用气人口	天 然 气 汽车加气站	地区名称
燃气汽车 Gas-Powered Automobiles	燃气损失量 Loss Amount	（户） Number of Household with Access to Gas (unit)	家庭用户 Household User	（万人） Population with Access to Gas (10,000 persons)	（座） Gas Stations for CNG-Fueled Motor Vehicles (unit)	Name of Regions
1087519.44	334226.41	180947860	176071871	41301.60	4267	全　　国
23215.00	74640.00	7244741	7155880	1475.63	97	北　京
29041.97	12213.09	6455018	5796394	1111.07	75	天　津
43433.07	19408.81	8425760	8073584	1737.18	199	河　北
25562.32	7621.76	4738515	4700019	1115.90	86	山　西
34121.07	1362.18	2516693	2409305	672.82	146	内　蒙　古
29372.65	8567.57	8062118	7900302	1654.30	162	辽　宁
42728.34	4866.08	4051499	3977033	859.32	224	吉　林
33407.56	2710.37	4318277	4237341	977.38	177	黑　龙　江
11953.43	31670.70	7694885	7534840	1874.95	23	上　海
54194.17	23377.99	14290765	13800595	3172.72	266	江　苏
49134.89	7249.59	6683678	6607723	1925.42	126	浙　江
46411.47	9801.88	6847586	6759141	1612.55	159	安　徽
14590.35	1201.87	2598158	2581413	799.99	68	福　建
8546.18	2921.76	3384323	3334188	910.28	39	江　西
99952.14	19968.64	15611696	15354231	3608.86	524	山　东
47425.87	16557.01	9236890	9075258	2220.16	197	河　南
50834.15	14818.79	8275517	8172944	1862.16	200	湖　北
20965.12	7094.66	4995526	4936747	1245.37	97	湖　南
45543.46	4863.06	10455005	10291939	3307.21	119	广　东
8863.68	618.76	2516692	2491667	703.91	33	广　西
13133.07	593.45	1018263	1009390	278.91	43	海　南
55363.75	14401.39	7587239	7236375	1489.53	109	重　庆
87062.03	20875.43	14457247	13480654	2577.82	250	四　川
6101.62	1103.14	2066938	2041737	495.20	26	贵　州
10916.85	556.28	2233187	2225599	563.07	38	云　南
	213.45	119560	119000	35.70		西　藏
30497.79	10584.82	6666986	6549771	1221.37	163	陕　西
24763.69	977.95	2258351	2221018	524.76	104	甘　肃
14350.87	6206.44	632592	614271	177.11	35	青　海
21307.14	1649.57	1334931	1319140	267.99	92	宁　夏
89819.39	4665.83	3752458	3660909	730.01	330	新　疆
14906.35	864.09	416766	403463	92.95	60	新疆生产 建设兵团

1-8-4 城市液化石油气(2020年)

地区名称 Name of Regions	储气能力 (吨) Gas Storage Capacity (ton)	供气管道长度 (公里) Length of Gas Supply Pipeline (km)	供气总量(吨)		
			合 计 Total	销售气量 Quantity Sold	居民家庭 Households
全　国　**National Total**	**960757.04**	**4010.00**	**8337109.06**	**8291148.13**	**4786679.10**
北　京　Beijing	18428.53	229.03	318293.00	292445.00	106695.00
天　津　Tianjin	4901.30		60575.60	59891.63	39142.95
河　北　Hebei	14758.76	140.12	80268.04	79789.18	55680.90
山　西　Shanxi	6933.68	13.40	57163.69	56586.90	43186.38
内 蒙 古　Inner Mongolia	7694.68	120.04	57053.46	56283.00	44240.90
辽　宁　Liaoning	39448.60	295.84	549329.18	548956.03	116119.28
吉　林　Jilin	19282.44	29.52	141174.35	140498.24	44377.29
黑 龙 江　Heilongjiang	11265.50	19.23	146843.88	145472.51	83686.29
上　海　Shanghai	21130.00	263.23	268425.20	272198.74	153474.19
江　苏　Jiangsu	61373.39	149.80	525334.53	524395.74	314540.54
浙　江　Zhejiang	42263.39	265.04	839084.15	836917.43	576886.46
安　徽　Anhui	16277.45	243.19	157718.65	156805.05	87247.55
福　建　Fujian	16879.30	194.79	290478.34	290223.72	177499.36
江　西　Jiangxi	19632.10	142.09	212089.19	210271.81	165542.20
山　东　Shandong	37056.53	2.70	291444.06	289910.72	188377.05
河　南　Henan	13190.00	14.20	165378.08	164055.20	138685.70
湖　北　Hubei	33894.43	129.97	271929.93	270388.21	182019.98
湖　南　Hunan	31859.70	41.60	252371.33	251169.99	198248.79
广　东　Guangdong	297073.48	868.75	2557229.02	2554763.06	1413010.86
广　西　Guangxi	144754.43	2.09	317493.81	316638.40	233802.85
海　南　Hainan	3687.65		77971.38	77949.17	70578.83
重　庆　Chongqing	7317.71		57149.64	56845.61	37258.04
四　川　Sichuan	19735.81	710.17	204997.88	202923.37	93579.07
贵　州　Guizhou	16319.39		117404.36	117200.45	49023.02
云　南　Yunnan	25514.39	18.13	150339.48	149876.38	56964.86
西　藏　Tibet	887.50	1.40	8595.67	8588.05	7868.05
陕　西　Shaanxi	10264.20	2.67	37063.65	36777.18	23955.92
甘　肃　Gansu	8650.80	110.00	33979.27	33751.26	24280.12
青　海　Qinghai	2926.70		17192.73	17016.93	15557.68
宁　夏　Ningxia	3740.00	0.82	14775.76	14716.26	6155.06
新　疆　Xinjiang	3420.20	2.18	54665.75	54570.91	35721.93
新疆生产 建设兵团　Xinjiang Production and Construction Corps	195.00		3296.00	3272.00	3272.00

Urban LPG Supply(2020)

Total Gas Supplied (ton)		用气户数 （户） Number of Household with Access to Gas (unit)	家庭用户 Household User	用气人口 （万人） Population with Access to Gas (10,000 persons)	液化石油气 汽车加气站 （座） Gas Stations for LPG- Fueled Motor Vehicles (unit)	地区名称 Name of Regions
燃气汽车 Gas-Powered Automobiles	燃气损失量 Loss Amount					
250229.41	45960.93	40237196	36873103	10767.47	332	全　　国
	25848.00	1880554	1866158	440.77		北　　京
	683.97	335964	330565	63.37		天　　津
	478.86	679037	615871	179.49		河　　北
	576.79	252677	197445	53.55	6	山　　西
2974.00	770.46	542064	477553	209.71	21	内　蒙　古
72021.70	373.15	1157045	1024067	286.77	50	辽　　宁
5217.00	676.11	753758	663324	229.89	25	吉　　林
33073.76	1371.37	940118	864899	274.53	44	黑　龙　江
20141.70	-3773.54	2305514	2223116	553.19	17	上　　海
14144.65	938.79	2058367	1811405	362.42	7	江　　苏
10489.01	2166.72	4482379	3798243	907.60	14	浙　　江
267.00	913.60	607586	519869	156.22	2	安　　徽
252.00	254.62	1530701	1445525	571.68	1	福　　建
10.00	1817.38	1251498	1182393	377.41	1	江　　西
22824.61	1533.34	1397638	1244635	352.67	54	山　　东
5629.00	1322.88	960330	908429	374.12	26	河　　南
3208.30	1541.72	1613017	1499693	384.77	6	湖　　北
2051.00	1201.34	1634534	1585172	463.95	7	湖　　南
52379.98	2465.96	9692866	9146665	2965.83	17	广　　东
	855.41	1999635	1881607	524.04		广　　西
	22.21	482612	478853	69.23		海　　南
	304.03	217137	160778	58.88		重　　庆
1566.00	2074.51	509737	399713	112.54	10	四　　川
	203.91	926786	787924	296.54		贵　　州
	463.10	921560	789848	244.99	2	云　　南
712.00	7.62	70224	68009	27.60	6	西　　藏
290.00	286.47	266341	207571	55.36	3	陕　　西
817.00	228.01	252288	224934	75.67	8	甘　　肃
420.80	175.80	54037	52231	25.16	2	青　　海
	59.50	143871	134599	24.47		宁　　夏
1739.90	94.84	285871	250559	39.07	3	新　　疆
	24.00	31450	31450	5.98		新疆生产 建设兵团

1-9-1　全国历年城市集中供热情况

年 份 Year	供 热 能 力 Heating Capacity		供 热 总 量 Total Heat Supplied	
	蒸 汽 （吨/小时） Steam (ton/hour)	热 水 （兆瓦） Hot Water (mega watts)	蒸 汽 （万吉焦） Steam (10,000 gigajoules)	热 水 （万吉焦） Hot Water (10,000 gigajoules)
1981	754	440	641	183
1982	883	718	627	241
1983	965	987	650	332
1984	1421	1222	996	454
1985	1406	1360	896	521
1986	9630	36103	3467	2704
1987	16258	27601	6669	3650
1988	18550	32746	5978	4848
1989	20177	25987	6782	4334
1990	20341	20128	7117	21658
1991	21495	29663	8195	21065
1992	25491	45386	9267	26670
1993	31079	48437	10633	29036
1994	34848	52466	10335	32056
1995	67601	117286	16414	75161
1996	62316	103960	17615	56307
1997	65207	69539	20604	62661
1998	66427	71720	17463	64684
1999	70146	80591	22169	69771
2000	74148	97417	23828	83321
2001	72242	126249	37655	100192
2002	83346	148579	57438	122728
2003	92590	171472	59136	128950
2004	98262	174442	69447	125194
2005	106723	197976	71493	139542
2006	95204	217699	67794	148011
2007	94009	224660	66374	158641
2008	94454	305695	69082	187467
2009	93193	286106	63137	200051
2010	105084	315717	66397	224716
2011	85273	338742	51777	229245
2012	86452	365278	51609	243818
2013	84362	403542	53242	266462
2014	84664	447068	55614	276546
2015	80699	472556	49703	302110
2016	78307	493254	41501	318044
2017	98328	647827	57985	310300
2018	92322	578244	57731	323665
2019	100943	550530	65067	327475
2020	103471	566181	65054	345004

注：1981年至1995年热水供热能力计量单位为兆瓦/小时；1981年至2000年蒸汽供热总量计量单位为万吨。

National Urban Centralized Heating in Past Years

蒸　汽 Steam	热　水 Hot Water	集中供热面积（万平方米）Heated Area (10,000 m²)	年份 Year
79	280	1167	1981
37	491	1451	1982
67	586	1841	1983
71	761	2445	1984
76	954	2742	1985
183	1335	9907	1986
163	1576	15282	1987
209	2193	13883	1988
401	2678	19386	1989
157	3100	21263	1990
656	3952	27651	1991
362	4230	32832	1992
532	5161	44164	1993
670	6399	50992	1994
909	8456	64645	1995
9577	24012	73433	1996
7054	25446	80755	1997
6933	27375	86540	1998
7733	30506	96775	1999
7963	35819	110766	2000
9183	43926	146329	2001
10139	48601	155567	2002
11939	58028	188956	2003
12775	64263	216266	2004
14772	71338	252056	2005
14012	79943	265853	2006
14116	88870	300591	2007
16045	104551	348948	2008
14317	110490	379574	2009
15122	124051	435668	2010
13381	133957	473784	2011
12690	147390	518368	2012
12259	165877	571677	2013
12476	174708	611246	2014
11692	192721	672205	2015
12180	201390	738663	2016
	276288	830858	2017
	371120	878050	2018
	392917	925137	2019
	425982	988209	2020

Note: Heating capacity through hot water from 1981 to 1995 is measured with the unit of megawatts/hour; Heating capacity through steam from 1981 to 2000 is measured with the unit of 10,000 tons.

1-9-2 城市集中供热(2020年)

地区名称 Name of Regions	蒸汽 Steam						供热能力 (兆瓦) Heating Capacity (mega watts)	热电厂供热 Heating by Co-Generation
	供热能力 (吨/小时) Heating Capacity (ton/hour)	热电厂供热 Heating by Co-Generation	锅炉房供热 Heating by Boilers	供热总量 (万吉焦) Total Heat Supplied (10,000 gcal)	热电厂供热 Heating by Co-Generation	锅炉房供热 Heating by Boilers		
全　国　National Total	103471	91597	10843	65054	57870	6512	566181	288168
北　京　Beijing							49656	9425
天　津　Tianjin	1875	1785	90	1014	983	31	31269	10722
河　北　Hebei	5813	5482	311	6977	6869	170	48237	27660
山　西　Shanxi	18029	17359	595	10682	10391	272	23614	16449
内　蒙　古　Inner Mongolia	3131	3061	70	2000	1964	36	49223	36713
辽　宁　Liaoning	16911	15373	1576	11414	10768	567	71732	25355
吉　林　Jilin	1958	1793	165	1073	1025	48	46265	19542
黑　龙　江　Heilongjiang	4901	4302	599	2498	2177	321	55185	31906
江　苏　Jiangsu							2	
安　徽　Anhui	2745	2425	320	3461	1475	1986	280	280
山　东　Shandong	27649	24635	2671	13849	12692	1010	68642	50871
河　南　Henan	6008	4642	856	3472	2637	413	24326	19929
湖　北　Hubei	1484	1434		1000	993		220	100
四　川　Sichuan								
贵　州　Guizhou							309	
云　南　Yunan							361	
西　藏　Tibet							64	64
陕　西　Shaanxi	6738	3338	3331	3489	2171	1308	25145	7646
甘　肃　Gansu	1100	1100		648	648		19398	8199
青　海　Qinghai							4852	4082
宁　夏　Ningxia	2004	2004		993	993		7766	7278
新　疆　Xinjiang	675	675		430	430		34967	11834
新疆生产建设兵团　Xinjiang Production and Construction Corps	2450	2190	260	2054	1655	350	4667	113

Urban Central Heating(2020)

| 热水　Hot Water | | | | 管道长度(公里)
Length of Pipelines
(km) | 一级管网
First Class | 二级管网
Second Class | 供热面积(万平方米)
Heated Area
(10,000 m²) | 住宅
Housing | 公共建筑
Public Building | 地区名称
Name of Regions |
锅炉房供热 Heating By Boilers	供热总量(万吉焦) Total Heat Supplied (10,000gcal)	热电厂供热 Heating by Co-Generation	锅炉房供热 Heating by Boilers							
206515	345004	195412	120804	425982	111303	314679	988209	742971	215577	全　国
	19189	5814		63740	4067	59673	65935	45486	20449	北　京
20547	17125	6765	10360	33687	8594	25093	55336	42826	12510	天　津
10633	28120	17466	5524	41990	9580	32410	87793	69584	15909	河　北
6502	16072	12698	3010	22412	7946	14466	74890	55291	15933	山　西
11420	33659	23961	9049	23518	6754	16763	63312	41656	16279	内蒙古
42803	51777	18545	30830	58834	14050	44784	132767	97986	31314	辽　宁
25420	27481	15999	11000	31996	7749	24247	67591	47517	17835	吉　林
22634	42887	25292	16857	21135	5174	15962	82604	58430	23698	黑龙江
2	3		3	0		0	4	4		江　苏
	5	5		741	722	19	2576	1516	1030	安　徽
15939	38698	29214	8544	79113	26189	52924	159346	134272	24643	山　东
2049	16052	13228	1319	12911	7223	5689	55995	47572	6605	河　南
	36	3		580	335	245	1759	1441	13	湖　北
				25		25				四　川
226	50		8	39	26	14	137	72	64	贵　州
32	76		25	460	145	315	110	50	60	云　南
	120	120		300	40	260	182	54	128	西　藏
13859	12144	3199	7355	4117	2772	1345	41685	34274	4492	陕　西
10721	12319	5334	6735	6285	1849	4435	27868	19343	6675	甘　肃
770	5411	4191	1220	2113	1238	874	10304	7084	3143	青　海
88	5518	5274	13	7132	2280	4852	14523	11207	2961	宁　夏
22621	17550	8254	8805	12990	3909	9081	38598	24318	10425	新　疆
249	712	49	147	1865	661	1204	4895	2990	1412	新疆生产建设兵团

三、居民出行数据
Data by Residents Travel

1-10-1　全国历年城市轨道交通情况
National Urban Rail Transit System in Past Years

年 份 Year	建成轨道交通 的城市个数 （个） Number of Cities with Completed Rail Transport Lines (unit)	建成轨道交通 线路长度 （公里） The Length of Completed Rail Transport Lines (km)	正在建设轨道 交通的城市个数 （个） Number of Cities with Rail Transport Lines under Construction (unit)	正在建设轨道 交通线路长度 （公里） The Length of Rail Transport Lines under Construction (km)
1978	1	23		
1979	1	23		
1980	1	23		
1981	1	23		
1982	1	23		
1983	1	23		
1984	2	47		
1985	2	47		
1986	2	47		
1987	2	47		
1988	2	47		
1989	2	47		
1990	2	47		
1991	2	47		
1992	2	47		
1993	2	47		
1994	2	47		
1995	3	63		

注：1.2000年及以前年份，大连、鞍山、长春、哈尔滨的有轨电车没有统计在内。

2.截至2020年底，在国务院已批复轨道交通建设规划的城市中，除包头市开工后停建外，其余城市已经全部开始建设或建成轨道交通线路。未含在以上城市名单中的昆山市、连云港市、淮安市、三亚市、蒙自市、天水市、嘉兴市、海宁市、湘潭市、都江堰市的上海市轨道交通11号线北段、连云港市域列车、淮安市现代有轨电车一期工程、三亚有轨电车示范线、滇南中心城市群现代有轨电车示范线项目、天水市有轨电车示范线工程(一期)、嘉兴市现代有轨电车T1线一期工程、嘉兴市现代有轨电车T2线一期工程(月河北站-环城南路站)、杭海城际铁路、长株潭城际轨道交通西环线一期(湘潭段)、万达文化旅游城交通配套项目-都江堰M-TR旅游客运专线工程PPP项目按城市轨道交通统计在内。

Note:1.For the year 2000 and before, the streetcar systems in Dalian, Anshan, Changchun and Harbin City were not included when collecting data on the number of cities with completed rail transport lines and the length of lines.

2.By the end of 2020, construction of rail transit lines in the cities whose rail transit construction plans have been approved by the State Council except the cities of Baotou has either started or completed. Data of Kunshan,Lianyungang,Huaian,Sanya,Mengzi,Tianshui,Jiaxing,Haining,Xiangtan and Dujiangyan which are not on the list of the cities, are included in the urban rail transit statistics.

1-10-1　续表　continued

年 份 Year	建成轨道交通 的城市个数 （个） Number of Cities with Completed Rail Transport Lines (unit)	建成轨道交通 线路长度 （公里） The Length of Completed Rail Transport Lines (km)	正在建设轨道 交通的城市个数 （个） Number of Cities with Rail Transport Lines under Construction (unit)	正在建设轨道 交通线路长度 （公里） The Length of Rail Transport Lines under Construction (km)
1996	3	63		
1997	3	63		
1998	4	81		
1999	4	81		
2000	4	117		
2001	5	172		
2002	5	200		
2003	5	347		
2004	7	400		
2005	10	444		
2006	10	621		
2007	10	775		
2008	10	855		
2009	10	838.88	28	1991.36
2010	12	1428.87	28	1741.07
2011	12	1672.42	28	1891.29
2012	16	2005.53	29	2060.43
2013	16	2213.28	35	2760.38
2014	22	2714.79	36	3004.37
2015	24	3069.23	38	3994.15
2016	30	3586.34	39	4870.18
2017	32	4594.26	50	4913.56
2018	34	5141.05	50	5400.25
2019	41	6058.90	49	5594.08
2020	42	7597.94	45	5093.55

1-10-2 城市轨道交通(建成)(2020年)

地区名称 Name of Regions	线路长度(公里) Length of Lines (km)								按敷设方式 by Ways of Laying		
	合计 Total	地铁 Subway	轻轨 Light Rail	单轨 Monorail	有轨 Cable Car	磁浮 Maglev	快轨 Fast Track	APM APM	地面线 Surface Lines	地下线 Underground Lines	高架线 Elevated Lines
全 国 **National Total**	**7597.94**	**6651.11**	**290.06**	**78.52**	**393.92**	**48.45**	**135.88**		**579.71**	**5445.42**	**1573.81**
北 京 Beijing	727.00	665.00			62.00				74.00	494.00	160.00
天 津 Tianjin	238.85	178.74	52.25		7.86				16.82	161.97	60.06
河 北 Hebei	61.60	61.60								61.60	
山 西 Shanxi	23.65	23.65								23.65	
内 蒙 古 Inner Mongolia	49.03	49.03							0.34	45.84	2.85
辽 宁 Liaoning	336.35	145.36	127.28		63.71				119.88	153.53	62.94
吉 林 Jilin	100.10	38.60	61.50						20.20	43.00	36.90
黑 龙 江 Heilongjiang	31.83	31.83								31.83	
上 海 Shanghai	719.37	683.18			6.29	29.90			16.40	436.71	266.26
江 苏 Jiangsu	834.75	718.81			80.84		35.10		118.37	511.46	204.92
浙 江 Zhejiang	507.78	454.27					53.51		3.23	391.16	113.39
安 徽 Anhui	112.49	112.49								108.26	4.23
福 建 Fujian	131.86	131.86							1.60	127.46	2.80
江 西 Jiangxi	88.71	88.71								88.71	
山 东 Shandong	302.67	293.90			8.77				10.89	171.76	120.02
河 南 Henan	215.52	215.52							1.27	198.22	16.03
湖 北 Hubei	388.01	338.88			49.13				44.45	263.28	80.28
湖 南 Hunan	161.19	142.64					18.55		0.22	141.29	19.68
广 东 Guangdong	1028.01	1002.09			25.92				47.70	869.91	110.40
广 西 Guangxi	108.06	108.06								108.06	
海 南 Hainan	8.37				8.37				8.02		0.35
重 庆 Chongqing	334.08	220.44	19.72	78.52	15.40				21.64	179.12	133.32
四 川 Sichuan	601.97	515.40			39.30		47.27		41.59	452.55	107.83
贵 州 Guizhou	35.11	35.11							2.36	29.02	3.73
云 南 Yunnan	152.86	139.46			13.40				15.70	115.15	22.01
西 藏 Tibet											
陕 西 Shaanxi	232.17	202.86	29.31						2.10	184.26	45.81
甘 肃 Gansu	38.93	26.00			12.93				12.93	26.00	
青 海 Qinghai											
宁 夏 Ningxia											
新 疆 Xinjiang	27.62	27.62								27.62	
新疆生产 建设兵团 Xinjiang Production and Construction Corps											

Urban Rail Transit System (Completed)(2020)

合计 Total	地面站 Surface Lines	地下站 Underground Lines	高架站 Elevated Lines	换乘站数(个) Number of Transfer Stations (unit)	合计 Total	地铁 Subway	轻轨 Light Rail	单轨 Monorail	有轨 Cable Car	磁浮 Maglev	快轨 Fast Track	APM APM	地区名称 Name of Regions
5019	539	3773	707	1303	39603	37882	380	492	685	32	132		全　国
428	42	318	68	132	6735	6606			129				北　京
181	20	130	31	42	1210	1034	152		24				天　津
50		50		5	408	408							河　北
23		23		7	144	144							山　西
44	1	40	3	10	312	312							内 蒙 古
258	122	115	21	54	907	746	128		33				辽　宁
94	21	37	36	16									吉　林
28		28		9	186	186							黑 龙 江
422	12	299	111	140	6174	6116			44	14			上　海
530	83	373	74	95	3459	3365			91			3	江　苏
304	2	246	56	57	2900	2868					32		浙　江
100		96	4	21	732	732							安　徽
103		102	1	27	393	393							福　建
74		74		17	630	630							江　西
153	12	97	44	46	1053	1046			7				山　东
152		145	7	59	239	239							河　南
291	60	180	51	89	2342	2284			58				湖　北
114		110	4	38	888	870				18			湖　南
629	62	517	50	165	3859	3784			75				广　东
82		82		21	104	104							广　西
15	15				11				11				海　南
217	19	110	88	61	1701	1094	100	492	15				重　庆
381	38	313	30	101	4575	4298			180		97		四　川
25	2	20	3	6	34	34							贵　州
107	15	85	7	31	143	125			18				云　南
													西　藏
155	1	136	18	44	276	276							陕　西
32	12	20		5	26	26							甘　肃
													青　海
													宁　夏
27		27		5	162	162							新　疆
													新疆生产建设兵团

1-10-3　城市轨道交通(在建)(2020年)

地区名称 Name of Regions	线路长度(公里) Length of Lines (km)										
	合计 Total	地铁 Subway	轻轨 Light Rail	单轨 Monorail	有轨 Cable Car	磁浮 Maglev	快轨 Fast Track	APM APM	按敷设方式 by Ways of Laying		
									地面线 Surface Lines	地下线 Underground Lines	高架线 Elevated Lines
全　国　National Total	5093.55	4696.77	199.44	46.25	33.86	1.20	116.03		142.80	4193.73	757.02
北　京　Beijing	325.22	324.02				1.20				260.62	64.60
天　津　Tianjin	219.64	219.64							1.40	193.34	24.90
河　北　Hebei	17.80	17.80								17.80	
山　西　Shanxi	28.74	28.74								28.74	
内 蒙 古　Inner Mongolia											
辽　宁　Liaoning	244.92	244.92							18.22	177.87	48.83
吉　林　Jilin	97.23	87.36	9.87							92.74	4.49
黑 龙 江　Heilongjiang	60.88	60.88								60.88	
上　海　Shanghai	171.63	171.63								140.03	31.60
江　苏　Jiangsu	574.73	574.73							2.80	525.49	46.44
浙　江　Zhejiang	567.32	322.08	115.43		13.78		116.03		28.22	357.36	181.74
安　徽　Anhui	230.65	184.40		46.25					0.31	178.55	51.79
福　建　Fujian	249.97	249.97							2.14	197.27	50.56
江　西　Jiangxi	39.60	39.60							0.20	33.91	5.49
山　东　Shandong	205.60	205.60							0.23	203.53	1.84
河　南　Henan	239.58	165.44	74.14						1.59	213.00	24.99
湖　北　Hubei	228.14	228.14							2.70	178.28	47.16
湖　南　Hunan	96.36	96.36							0.42	76.72	19.22
广　东　Guangdong	685.50	685.50							61.76	578.73	45.01
广　西　Guangxi	24.28	24.28								24.28	
海　南　Hainan											
重　庆　Chongqing	202.51	202.51							3.81	133.34	65.36
四　川　Sichuan	198.30	178.22			20.08				17.36	170.71	10.23
贵　州　Guizhou	113.95	113.95							1.64	101.74	10.57
云　南　Yunnan	48.33	48.33								48.33	
西　藏　Tibet											
陕　西　Shaanxi	154.23	154.23								132.03	22.20
甘　肃　Gansu	9.06	9.06								9.06	
青　海　Qinghai											
宁　夏　Ningxia											
新　疆　Xinjiang	59.38	59.38								59.38	
新疆生产 建设兵团　Xinjiang Production and Construction Corps											

Urban Rail Transit System(Under Construction) (2020)

车站数(个) Number of Stations (unit)				换乘站数(个) Number of Transfer Stations (unit)	配置车辆数(辆) Number of Vehicles in Service (unit)								地区名称 Name of Regions
合计 Total	地面站 Surface Lines	地下站 Under-ground Lines	高架站 Elevated Lines		合计 Total	地铁 Subway	轻轨 Light Rail	单轨 Monorail	有轨 Cable Car	磁浮 Maglev	快轨 Fast Track	APM APM	
3144	67	2816	261	1063	23182	22249	433	240	39	60	161		全　　国
164		144	20	89	4010	3950				60			北　　京
166	1	154	11	70	1086	1086							天　　津
14		14			78	78							河　　北
24		24		7	162	162							山　　西
													内　蒙　古
149	3	131	15	52	1214	1214							辽　　宁
77		72	5	29									吉　　林
50		50		15	540	540							黑　龙　江
98		93	5	31	1072	1072							上　　海
397	2	390	5	144	3804	3804							江　　苏
278	21	203	54	63	2330	1784	368		17		161		浙　　江
161		121	40	43	1404	1164		240					安　　徽
125		113	12	51	616	616							福　　建
29		25	4	7	276	276							江　　西
140	1	137	2	57	798	798							山　　东
169	8	160	1	60	497	432	65						河　　南
119		103	16	33	1310	1310							湖　　北
62	1	53	8	14	634	634							湖　　南
370		358	12	95	91	91							广　　东
20		20		4	24	24							广　　西
													海　　南
117	4	85	28	46	1132	1132							重　　庆
145	26	110	9	72	1496	1474			22				四　　川
74		70	4	19	113	113							贵　　州
40		40		14	59	59							云　　南
													西　　藏
102		92	10	24									陕　　西
9		9		5	10	10							甘　　肃
													青　　海
													宁　　夏
45		45		19	426	426							新　　疆
													新疆生产建设兵团

1-11-1　全国历年城市道路和桥梁情况
National Urban Road and Bridge in Past Years

年 份 Year	道路长度 (公里) Length of Roads (km)	道路面积 (万平方米) Surface Area of Roads (10,000 m²)	防洪堤长度 (公里) Length of Flood Control Dikes (km)	人均城市道路面积 (平方米) Urban Road Surface Area Per Capita (m²)
1978	26966	22539	3443	2.93
1979	28391	24069	3670	2.85
1980	29485	25255	4342	2.82
1981	30277	26022	4446	1.81
1982	31934	27976	5201	1.96
1983	33934	29962	5577	1.88
1984	36410	33019	6170	1.84
1985	38282	35872	5998	1.72
1986	71886	69856	9952	3.05
1987	78453	77885	10732	3.10
1988	88634	91355	12894	3.10
1989	96078	100591	14506	3.22
1990	94820	101721	15500	3.13
1991	88791	99135	13892	3.35
1992	96689	110526	16015	3.59
1993	104897	124866	16729	3.70
1994	111058	137602	16575	3.84
1995	130308	164886	18885	4.36

注：1.自2006年起，人均城市道路面积按城区人口和城区暂住人口合计为分母计算，括号内为与往年同口径数据。
　　2.自2013年开始，不再统计防洪堤长度。

Note: 1.Since 2006, urban road surface per capita has been calculated based on denominator which combines both permanent and
　　temporary residents in urban areas, and the data in brackets are the same index calculated by the method of past years.
　　2.Starting from 2013, the data on the length of flood prevention dyke has been unavailable.

1-11-1 续表   continued

年 份 Year	道路长度 (公里) Length of Roads (km)	道路面积 (万平方米) Surface Area of Roads (10,000 m^2)	防洪堤长度 (公里) Length of Flood Control Dikes (km)	人均城市道路面积 (平方米) Urban Road Surface Area Per Capita (m^2)
1996	132583	179871	18475	4.96
1997	138610	192165	18880	5.22
1998	145163	206136	19550	5.51
1999	152385	222158	19842	5.91
2000	159617	237849	20981	6.13
2001	176016	249431	23798	6.98
2002	191399	277179	25503	7.87
2003	208052	315645	29426	9.34
2004	222964	352955	29515	10.34
2005	247015	392166	41269	10.92
2006	241351	411449	38820	11.04(12.36)
2007	246172	423662	32274	11.43
2008	259740	452433	33147	12.21
2009	269141	481947	34698	12.79
2010	294443	521322	36153	13.21
2011	308897	562523	35051	13.75
2012	327081	607449	33926	14.39
2013	336304	644155		14.87
2014	352333	683028		15.34
2015	364978	717675		15.60
2016	382454	753819		15.80
2017	397830	788853		16.05
2018	432231	854268		16.70
2019	459304	909791		17.36
2020	492650	969803		18.04

1-11-2 城市道路和桥梁（2020年）

地区名称 Name of Regions	道路长度 （公里） Length of Roads （km）	建成区 in Built District	道路面积 （万平方米） Surface Area of Roads （10,000 m²）	人行道 面 积 Surface Area of Sidewalks	建成区 in Built District
全 国 National Total	**492650.37**	**429050.34**	**969802.54**	**211589.82**	**861449.60**
北 京 Beijing	8405.60		14702.00	2537.00	
天 津 Tianjin	9233.87	7922.18	17510.39	4081.30	14743.44
河 北 Hebei	18766.58	18193.87	41074.02	9966.49	39191.96
山 西 Shanxi	9803.02	9123.94	22325.23	5099.54	19935.59
内 蒙 古 Inner Mongolia	10505.67	9661.59	22074.23	5454.69	21566.33
辽 宁 Liaoning	21481.85	18477.96	36597.86	8524.96	32651.52
吉 林 Jilin	10952.30	9420.12	19116.00	4890.23	17072.73
黑 龙 江 Heilongjiang	13712.57	12994.38	22075.68	4780.88	21053.66
上 海 Shanghai	5536.00	5536.00	11551.00	2873.00	11551.00
江 苏 Jiangsu	50860.81	42629.36	90570.25	14322.10	73868.28
浙 江 Zhejiang	27573.61	23305.56	54047.75	11574.61	46771.46
安 徽 Anhui	17750.49	16835.02	43286.17	9285.57	41090.19
福 建 Fujian	14386.10	12964.42	26167.50	4919.01	22596.41
江 西 Jiangxi	12655.92	11775.65	26279.41	5803.79	23613.03
山 东 Shandong	49986.19	43537.99	102269.10	20347.46	90681.67
河 南 Henan	16294.57	15542.54	41039.33	9616.76	38872.46
湖 北 Hubei	22992.18	22332.89	43142.81	10581.27	42069.98
湖 南 Hunan	15241.53	13576.52	34645.25	8352.66	31582.60
广 东 Guangdong	49374.02	40616.30	84013.94	17547.65	74616.18
广 西 Guangxi	14918.66	13510.92	30186.29	5120.13	26636.51
海 南 Hainan	4516.56	4449.32	6301.47	2122.50	4661.17
重 庆 Chongqing	10872.67	10267.98	23592.88	7142.33	22269.08
四 川 Sichuan	25537.86	23775.62	50907.26	13476.85	48533.93
贵 州 Guizhou	9261.56	7947.88	17779.92	4528.33	15487.85
云 南 Yunnan	8242.17	7970.11	17078.86	3814.63	16623.66
西 藏 Tibet	988.27	675.07	2076.84	709.45	1616.90
陕 西 Shaanxi	9478.91	7106.83	21660.13	4925.91	19572.12
甘 肃 Gansu	5986.65	5788.34	13130.44	3137.20	12369.87
青 海 Qinghai	1642.37	1420.93	4077.36	977.75	3756.35
宁 夏 Ningxia	2953.32	2798.46	8031.27	1406.87	7518.11
新 疆 Xinjiang	11270.58	7536.77	19889.92	3101.66	16520.05
新疆生产 建设兵团 Xinjiang Production and Construction Corps	1467.91	1355.82	2601.98	567.24	2355.51

Urban Roads and Bridges(2020)

桥梁数 （座） Number of Bridges (unit)	大桥及 特大桥 Great Bridge and Grand Bridge	立交桥 Inter-section	道路照明 灯盏数 （盏） Number of Road Lamps (unit)	安装路灯 道路长度 （公里） Length of The Road with Street Lamp (km)	地下综合 管廊长度 （公里） Length of The Utility Tunnel (km)	新建地下 综合管廊长度 （公里） Length of The New-bwit Utility Tunnel (km)	地区名称 Name of Regions
79752	**10103**	**5625**	**30485606**	**385958.13**	**6150.76**	**2690.68**	全　　国
2376	554	457	315285	8402.00	199.96	7.90	北　　京
1196	211	141	396388	8103.17	16.55	5.04	天　　津
1870	165	222	1068008	12624.63	325.21	93.56	河　　北
602	81	90	512740	6218.03	25.72	7.17	山　　西
505	97	65	618492	7029.35	67.43	26.83	内　蒙　古
1898	301	227	1354530	16062.06	102.60	2.84	辽　　宁
966	209	111	548217	7408.69	233.97	72.97	吉　　林
1233	270	263	712097	10824.08	25.84		黑　龙　江
2880	3	50	640886	5536.00	91.04	16.78	上　　海
16932	841	439	3662878	39464.40	277.18	82.45	江　　苏
12703	640	200	1831720	23062.62	151.48	243.90	浙　　江
2027	228	324	1117700	13589.68	91.58	94.75	安　　徽
1859	271	49	857760	8682.80	318.09	61.95	福　　建
1114	155	104	863564	9355.31	196.95	108.53	江　　西
5821	350	216	2119319	39443.27	750.91	152.41	山　　东
1624	128	198	1060950	13356.32	81.93	111.07	河　　南
2332	465	203	967971	19519.60	401.87	185.11	湖　　北
1311	432	104	863176	13160.91	172.16	18.78	湖　　南
8296	2552	972	3575478	49661.98	413.53	236.98	广　　东
1346	251	159	787613	8052.08	149.42	44.46	广　　西
229	15	11	193917	2509.50	61.13	56.44	海　　南
2173	343	306	830550	9484.36	36.12	18.50	重　　庆
3577	554	274	1768992	20886.80	743.16	402.58	四　　川
770	184	64	731940	4379.37	67.65	13.61	贵　　州
1374	347	93	671963	6830.91	304.94	69.41	云　　南
62	7	1	31970	672.82	51.30	8.32	西　　藏
840	141	80	832489	7011.80	496.78	356.98	陕　　西
697	128	88	392918	3988.08	54.18	35.38	甘　　肃
235	3	4	152244	1077.62	108.42		青　　海
222	10	8	245021	2352.95	47.50	4.70	宁　　夏
631	157	93	644970	6358.90	25.79	85.39	新　　疆
51	10	9	113860	848.04	60.37	65.89	新疆生产 建设兵团

四、环境卫生数据
Data by Environmental Health

1-12-1　全国历年城市排水和污水处理情况
National Urban Drainage and Wastewater Treatment in Past Years

年 份 Year	排水管道长度 (公里) Length of Drainage Pipelines (km)	污水年 排放量 (万立方米) Annual Quantity of Wastewater Discharged (10,000 m³)	污水处理厂 Wastewater Treatment Plant		污水年处理量 (万立方米) Annual Treatment Capacity (10,000 m³)	污水处理率 (%) Wastewater Treatment Rate (%)
			座数 (座) Number (unit)	处理能力 (万立方米/日) Treatment Capacity (10,000 m³/day)		
1978	19556	1494493	37	64		
1979	20432	1633266	36	66		
1980	21860	1950925	35	70		
1981	23183	1826460	39	85		
1982	24638	1852740	39	76		
1983	26448	2097290	39	90		
1984	28775	2253145	43	146		
1985	31556	2318480	51	154		
1986	42549	963965	64	177		
1987	47107	2490249	73	198		
1988	50678	2614897	69	197		
1989	54510	2611283	72	230		
1990	57787	2938980	80	277		
1991	61601	2997034	87	317	445355	14.86
1992	67672	3017731	100	366	521623	17.29
1993	75207	3113420	108	449	623163	20.02
1994	83647	3030082	139	540	518013	17.10
1995	110293	3502553	141	714	689686	19.69
1996	112812	3528472	309	1153	833446	23.62
1997	119739	3514011	307	1292	907928	25.84
1998	125943	3562912	398	1583	1053342	29.56
1999	134486	3556821	402	1767	1135532	31.93
2000	141758	3317957	427	2158	1135608	34.25

注：1978年至1995年污水处理厂座数及处理能力均为系统内数。

Note: Number of wastewater treatment plants and treatment capacity from 1978 to 1995 are limited to the statistical figure in the building sector.

续表　continued

年 份 Year	排水管道长度 (公里) Length of Drainage Pipelines (km)	污水年排放量 (万立方米) Annual Quantity of Wastewater Discharged (10,000 m³)	污水处理厂 Wastewater Treatment Plant		污水年处理量 (万立方米) Annual Treatment Capacity (10,000 m³)	污水处理率 (%) Wastewater Treatment Rate (%)
			座数 (座) Number (unit)	处理能力 (万立方米/日) Treatment Capacity (10,000 m³/day)		
2001	158128	3285850	452	3106	1196960	36.43
2002	173042	3375959	537	3578	1349377	39.97
2003	198645	3491616	612	4254	1479932	42.39
2004	218881	3564601	708	4912	1627966	45.67
2005	241056	3595162	792	5725	1867615	51.95
2006	261379	3625281	815	6366	2026224	55.67
2007	291933	3610118	883	7146	2269847	62.87
2008	315220	3648782	1018	8106	2560041	70.16
2009	343892	3712129	1214	9052	2793457	75.25
2010	369553	3786983	1444	10436	3117032	82.31
2011	414074	4037022	1588	11303	3376104	83.63
2012	439080	4167602	1670	11733	3437868	87.30
2013	464878	4274525	1736	12454	3818948	89.34
2014	511179	4453428	1807	13087	4016198	90.18
2015	539567	4666210	1944	14038	4288251	91.90
2016	576617	4803049	2039	14910	4487944	93.44
2017	630304	4923895	2209	15743	4654910	94.54
2018	683485	5211249	2321	16881	4976126	95.49
2019	743982	5546474	2471	17863	5369283	96.81
2020	802721	5713633	2618	19267	5572782	97.53

1-12-2 城市排水和污水处理(2020年)

地区名称 Name of Regions	污水排放量 (万立方米) Annual Quantity of Wastewater Discharged (10,000 m³)	排水管道长度 (公里) Length of Drainage Piplines (km)	污水管道 Sewers	雨水管道 Rainwater Drainage Pipeline	雨污合流管道 Combined Drainage Pipeline	建成区 in Built District	污水处理厂			
							座数 (座) Number of Wastewater Treatment Plant (unit)	二、三级处理 Secondary and Tertiary Treatment	处理能力 (万立方米/日) Treatment Capacity (10,000 m³/day)	二、三级处理 Secondary and Tertiary Treatment
全 国 National Total	5713633	802721	366833	334753	101134	683914	2618	2441	19267.1	18344.6
北 京 Beijing	191278	17943	8772	7290	1881	9503	70	70	687.9	687.9
天 津 Tianjin	112734	22338	10338	10725	1275	21910	44	44	338.5	338.5
河 北 Hebei	174010	20927	10542	10184	201	20216	93	91	680.1	665.5
山 西 Shanxi	96131	12250	5818	5163	1269	9737	48	40	343.1	301.7
内 蒙 古 Inner Mongolia	66193	14371	7878	5912	581	12756	41	39	236.6	234.1
辽 宁 Liaoning	313244	23483	6467	8205	8812	20240	131	96	1009.4	818.0
吉 林 Jilin	130519	13552	5522	6322	1708	10916	50	36	445.0	364.5
黑 龙 江 Heilongjiang	123582	13291	3999	4759	4533	12818	69	68	416.2	406.2
上 海 Shanghai	221504	17230	7944	8054	1233	17230	42	42	840.3	840.3
江 苏 Jiangsu	479860	88001	45997	34066	7938	71112	206	203	1480.9	1456.9
浙 江 Zhejiang	338632	52572	26919	23104	2549	43807	106	106	1173.9	1173.9
安 徽 Anhui	208669	35393	15432	17198	2763	32504	96	95	723.5	718.5
福 建 Fujian	143773	20114	9286	9731	1097	18034	55	54	428.5	423.5
江 西 Jiangxi	110960	20023	8676	7702	3645	18733	68	54	360.7	305.7
山 东 Shandong	341815	69864	30332	37159	2373	63932	218	218	1364.8	1364.8
河 南 Henan	194771	29222	12450	12690	4081	26924	110	106	890.3	846.3
湖 北 Hubei	287470	32641	12435	14441	5766	28369	101	98	868.7	848.4
湖 南 Hunan	243174	21665	7861	9292	4512	19571	92	76	741.5	658.0
广 东 Guangdong	830750	122541	58407	41560	22574	90719	320	303	2714.8	2602.9
广 西 Guangxi	153058	19174	6657	7862	4655	18791	63	62	452.1	444.1
海 南 Hainan	37156	6524	3008	2698	818	4018	25	22	118.9	111.8
重 庆 Chongqing	142252	23542	11223	10584	1735	21829	80	76	411.9	391.4
四 川 Sichuan	257678	42610	19328	18475	4806	34957	149	141	788.9	775.1
贵 州 Guizhou	93831	10394	5493	3991	910	9259	101	101	345.5	345.5
云 南 Yunnan	109046	16328	7756	6871	1701	14244	59	56	309.0	302.5
西 藏 Tibet	10037	861	301	272	288	569	9	6	28.7	10.2
陕 西 Shaanxi	131968	12402	5917	5231	1254	10068	57	47	415.4	364.4
甘 肃 Gansu	48396	8035	4211	2386	1438	7184	30	28	169.1	165.3
青 海 Qinghai	18465	3301	1695	1381	225	3165	14	13	61.8	60.3
宁 夏 Ningxia	28138	2297	500	442	1355	2167	23	20	118.6	108.1
新 疆 Xinjiang	64097	8510	4984	948	2578	7418	38	26	254.4	203.6
新疆生产建设兵团 Xinjiang Production and Construction Corps	10444	1325	687	55	582	1213	10	4	48.7	7.7

Urban Drainage and Wastewater Treatment(2020)

Wastewater Treatment Plant				其他污水处理设施 Other Wastewater Treatment Facilities		污水处理总量(万立方米) Total Quantity of Wastewater Treated	市政再生水 Recycled Water			地区名称 Name of Regions
处理量(万立方米) Quantity of Wastewater Treated (10,000 m³)	二、三级处理 Secondary and Tertiary Treated	干污泥产生量(吨) Quantity of Dry Sludge Produced (ton)	干污泥处置量(吨) Quantity of Dry Sludge Treated (ton)	处理能力(万立方米/日) Treatment Capacity (10,000 m³/day)	处理量(万立方米) Quantity of Wastewater Treated (10,000 m³)	(10,000 m³)	生产能力(万立方米/日) Recycled Water Production Capacity (10,000 m³/day)	利用量(万立方米) Annual Quantity of Wastewater Recycled and Reused (10,000 m³)	管道长度(公里) Length of Piplines (km)	
5472276	5213315	11627678	11160222	1138.0	100506	5572782	6095.2	1353832	14630	全　国
181252	181252	1656419	1655305	22.8	3455	184707	687.9	120133	2074	北　京
107704	107704	148352	148334	3.4	994	108698	171.2	35470	1988	天　津
171325	169237	407529	406142			171325	483.5	70861	744	河　北
95747	85301	249548	248531			95747	235.5	23820	512	山　西
64738	63864	216651	212454			64738	162.1	25305	1445	内 蒙 古
305222	249521	526327	525727	11.4	2486	307708	243.3	33767	287	辽　宁
127508	104048	200866	197903			127508	77.1	18561	76	吉　林
113893	110737	168446	162770	46.0	4726	118620	43.1	26773	104	黑 龙 江
213021	213021	484087	484087		1124	214144				上　海
432770	427733	1001562	984036	373.7	31840	464610	510.3	125539	938	江　苏
327630	327630	784583	784582	13.4	3165	330795	185.1	38754	261	浙　江
199819	198981	253727	253693	55.8	3482	203300	284.3	77948	164	安　徽
134279	132717	194923	194870	46.7	5400	139680	163.3	28980	100	福　建
107047	92269	100459	86702	5.4	1120	108167		24		江　西
335352	335352	730681	730432	11.7	521	335873	600.9	148312	1304	山　东
191456	180450	448460	355162	6.5	37	191493	337.3	72026	650	河　南
265161	258921	332955	332798	59.6	13598	278759	171.9	47695	20	湖　北
235758	210018	625640	612926	25.1	2045	237804	83.4	20422	136	湖　南
810532	772412	1141883	1141969	33.8	742	811273	862.2	280394	57	广　东
135809	133672	130602	133096	366.6	15708	151517	50.5	16736	2	广　西
36610	34230	64138	64138	0.5	56	36666	22.5	2461	140	海　南
139185	133615	243490	3703	2.9	457	139642	15.5	1486	77	重　庆
241109	236168	408349	389531	40.6	8488	249598	131.3	32023	81	四　川
91426	91426	109597	107696			91426	33.2	5200	17	贵　州
105397	103133	148873	141636	9.9	1062	106459	48.9	34628	591	云　南
9663	3556	2352	2344			9663	1.5	2	25	西　藏
127737	114817	411664	386865			127737	216.4	27550	344	陕　西
47031	46406	114011	94228			47031	58.5	5739	1059	甘　肃
17599	16958	24537	22068			17599	18.1	3088	80	青　海
27222	25001	54135	54130			27222	53.2	6769	473	宁　夏
62887	51221	190199	189733	2.0		62887	112.2	22526	859	新　疆
10385	1945	52632	52632	0.4		10385	31.0	840	24	新疆生产建设兵团

1-13-1　全国历年城市市容环境卫生情况

年 份 Year	生活垃圾　Domestic Garbage			
	清运量 (万吨) Quantity of Collected and Transported (10,000 ton)	无害化处理场(厂)座数 (座) Number of Harmless Treatment Plants/ Grounds (unit)	无害化处理能力 (吨／日) Harmless Treatment Capacity (ton/day)	无害化处理量 (万吨) Quantity of Harmlessly Treated (10,000 ton)
1979	2508	12	1937	
1980	3132	17	2107	215
1981	2606	30	3260	162
1982	3125	27	2847	190
1983	3452	28	3247	243
1984	3757	24	1578	188
1985	4477	14	2071	232
1986	5009	23	2396	70
1987	5398	23	2426	54
1988	5751	29	3254	75
1989	6292	37	4378	111
1990	6767	66	7010	212
1991	7636	169	29731	1239
1992	8262	371	71502	2829
1993	8791	499	124508	3945
1994	9952	609	130832	4782
1995	10671	932	183732	6014
1996	10825	574	155826	5568
1997	10982	635	180081	6292
1998	11302	655	201281	6783
1999	11415	696	237393	7232
2000	11819	660	210175	7255
2001	13470	741	224736	7840
2002	13650	651	215511	7404
2003	14857	575	219607	7545
2004	15509	559	238519	8089
2005	15577	471	256312	8051
2006	14841	419	258048	7873
2007	15215	458	279309	9438
2008	15438	509	315153	10307
2009	15734	567	356130	11220
2010	15805	628	387607	12318
2011	16395	677	409119	13090
2012	17081	701	446268	14490
2013	17239	765	492300	15394
2014	17860	818	533455	16394
2015	19142	890	576894	18013
2016	20362	940	621351	19674
2017	21521	1013	679889	21034
2018	22802	1091	766195	22565
2019	24206	1183	869875	24013
2020	23512	1287	963460	23452

注：1.1980年至1995年垃圾无害化处理厂,垃圾无害化处理量为垃圾加粪便。

2.自2006年起，生活垃圾填埋场的统计采用新的认定标准，生活垃圾无害化处理数据与往年不可比。

National Urban Environmental Sanitation in Past Years

粪便清运量 （万吨） Volume of Soil Collected and Transported （10,000 ton）	公共厕所 （座） Number of Latrine （unit）	市容环卫专用 车辆设备总数 （辆） Number of Vehicles and Equipment Designated for Municipal Environmental Sanitation （unit）	每 万 人 拥有公厕 （座） Number of Latrine per 10,000 Population （unit）	年 份 Year
2156	54180	5316		1979
1643	61927	6792		1980
1547	54280	7917	3.77	1981
1689	56929	9570	3.99	1982
1641	62904	10836	3.95	1983
1538	64178	11633	3.57	1984
1748	68631	13103	3.28	1985
2710	82746	19832	3.61	1986
2422	88949	21418	3.54	1987
2353	92823	22793	3.14	1988
2603	96536	25076	3.09	1989
2385	96677	25658	2.97	1990
2764	99972	27854	3.38	1991
3002	95136	30026	3.09	1992
3168	97653	32835	2.89	1993
3395	96234	34398	2.69	1994
3066	113461	39218	3.00	1995
2931	109570	40256	3.02	1996
2845	108812	41538	2.95	1997
2915	107947	42975	2.89	1998
2844	107064	44238	2.85	1999
2829	106471	44846	2.74	2000
2990	107656	50467	3.01	2001
3160	110836	52752	3.15	2002
3475	107949	56068	3.18	2003
3576	109629	60238	3.21	2004
3805	114917	64205	3.20	2005
2131	107331	66020	2.88(3.22)	2006
2506	112604	71609	3.04	2007
2331	115306	76400	3.12	2008
2141	118525	83756	3.15	2009
1951	119327	90414	3.02	2010
1963	120459	100340	2.95	2011
1812	121941	112157	2.89	2012
1682	122541	126552	2.83	2013
1552	124410	141431	2.79	2014
1437	126344	165725	2.75	2015
1299	129818	193942	2.72	2016
	136084	228019	2.77	2017
	147466	252484	2.88	2018
	153426	281558	2.93	2019
	165186	306422	3.07	2020

Note: 1.Quantity of garbage disposed harmlessly from 1980 to 1995 consists of quantity of garbage and soil.

2.Since 2006, treatment of domestic garbage through sanitary landfill has adopted new certification standard,so the datas of harmless treatmented garbage are not compared with the past years.

1-13-2　城市市容环境卫生(2020年)

地区名称 Name of Regions	道路清扫保洁面积 (万平方米) Surface Area of Roads Cleaned and Maintained (10,000 m²)	机械化 Mecha-nization	生活垃圾							
			清运量 (万吨) Collected and Trans-ported (10,000 ton)	处理量 (万吨) Volume of Treated (10,000 ton)	无害化处理厂(场)数(座) Number of Harmless Treatment Plants/Grounds (unit)	卫生填埋 Sanitary Landfill	焚烧 Inciner-ation	其他 other	无害化处理能力 (吨／日) Harmless Treatment Capacity (ton/day)	卫生填埋 Sanitary Landfill
全 国 National Total	975595	742539	23511.71	23492.68	1287	644	463	180	963460	337848
北 京 Beijing	16775	11471	797.52	797.52	41	9	12	20	33811	7491
天 津 Tianjin	13287	11644	306.54	306.54	19	4	11	4	20400	5100
河 北 Hebei	40082	36020	786.20	786.19	59	38	15	6	32173	11743
山 西 Shanxi	22495	18677	460.74	460.74	31	21	8	2	16783	9207
内 蒙 古 Inner Mongolia	26007	20892	387.69	387.38	29	25	4		12948	8798
辽 宁 Liaoning	47435	25823	993.26	993.26	41	29	8	4	32837	18737
吉 林 Jilin	18410	13749	464.17	464.17	34	21	10	3	18925	8795
黑 龙 江 Heilongjiang	27643	21338	497.63	496.96	45	34	7	4	22498	15003
上 海 Shanghai	18699	17458	868.09	868.09	23	5	10	8	40046	15350
江 苏 Jiangsu	69489	64219	1870.53	1870.53	85	25	46	14	83051	13715
浙 江 Zhejiang	53870	41695	1444.93	1444.93	76	16	43	17	76603	12133
安 徽 Anhui	42841	39632	660.70	660.70	49	16	22	11	32242	8182
福 建 Fujian	21399	15469	878.54	878.54	35	10	18	7	27759	4379
江 西 Jiangxi	25992	21274	527.53	527.53	29	12	14	3	23293	7690
山 东 Shandong	77847	64210	1673.94	1673.94	97	31	49	17	67636	14946
河 南 Henan	44651	36264	1130.20	1129.52	48	36	11	1	37366	20516
湖 北 Hubei	45008	34020	987.43	987.43	63	34	16	13	36597	14569
湖 南 Hunan	29576	25857	797.14	797.14	43	26	11	6	32355	15851
广 东 Guangdong	120331	65753	3102.55	3100.98	118	48	57	13	136593	43558
广 西 Guangxi	27715	17376	519.60	519.60	33	18	14	1	18972	7172
海 南 Hainan	9962	6764	253.64	253.64	12	6	4	2	8685	2310
重 庆 Chongqing	22755	15300	628.49	625.09	25	16	7	2	19449	8049
四 川 Sichuan	45041	33842	1136.57	1136.49	51	22	25	4	39444	13175
贵 州 Guizhou	16721	14472	358.48	350.77	35	14	14	7	18607	6347
云 南 Yunnan	19041	14050	487.51	487.45	36	24	11	1	17195	6145
西 藏 Tibet	2140	1164	62.34	62.11	11	10	1		3055	2355
陕 西 Shaanxi	20880	18444	550.01	549.64	33	28	4	1	18710	11699
甘 肃 Gansu	13580	11100	272.55	272.55	24	19	3	2	9593	5674
青 海 Qinghai	4015	2429	116.19	115.37	12	10		2	2140	1970
宁 夏 Ningxia	9376	6672	126.85	126.80	14	7	4	3	5936	2630
新 疆 Xinjiang	18658	12722	335.59	332.62	28	23	3	2	16455	14055
新疆生产建设兵团 Xinjiang Production and Construction Corps	3875	2739	28.59	28.48	8	7	1		1305	505

Urban Environmental Sanitation(2020)

Domestic Garbage						公共厕所 (座) Number of Latrines (unit)	三类 以上 Grade Ⅲ and Above	市容环卫 专用车辆 设备总数 (辆) Number of Vehicles and Equipment Designated for Munici- pal Environ- mental Sanitation (unit)	地区名称 Name of Regions
焚烧 Inciner- ation	其他 other	无害化 处理量 (万吨) Volume of Harmlessly Treated (10,000 ton)	卫生 填埋 Sanitary Landfill	焚烧 Inciner- ation	其他 other				
567804	**57808**	**23452.33**	**7771.54**	**14607.64**	**1073.15**	**165186**	**141279**	**306422**	**全　国**
18090	8230	797.52	111.64	507.21	178.66	6177	6177	13403	北　京
13550	1750	306.54	79.25	195.66	31.63	4269	3810	5568	天　津
18460	1970	786.19	309.70	447.24	29.25	6611	5383	12684	河　北
7298	278	460.74	284.15	167.30	9.29	3015	2057	7674	山　西
4150		387.38	268.59	118.79		7237	5050	6987	内 蒙 古
12580	1520	988.56	574.01	375.57	38.97	4664	3281	11196	辽　宁
9450	680	464.17	174.71	279.22	10.24	4319	3507	8003	吉　林
6452	1043	496.96	349.18	130.21	17.58	6048	3397	9144	黑 龙 江
19300	5396	868.09	69.95	682.11	116.02	5676	3799	10000	上　海
65420	3916	1870.53	196.61	1599.10	74.81	14961	13729	21351	江　苏
58630	5840	1444.93	294.71	1068.51	81.71	8631	7595	10316	浙　江
21510	2550	660.70	87.79	548.94	23.98	4769	4387	10354	安　徽
20800	2580	878.54	131.63	699.12	47.79	6344	5022	7342	福　建
15200	403	527.53	195.86	324.74	6.93	4331	4028	9695	江　西
49450	3240	1673.94	137.83	1476.73	59.38	8028	7470	21675	山　东
16800	50	1129.52	683.05	445.23	1.24	10890	10279	18459	河　南
16205	5824	987.43	423.47	479.12	84.84	6197	5588	14587	湖　北
14619	1885	797.14	348.42	415.64	33.08	4380	3205	7069	湖　南
87416	5620	3100.98	896.41	2108.37	96.21	12288	11788	28247	广　东
11650	150	519.60	251.63	262.07	5.89	1934	1772	10164	广　西
5875	500	253.64	80.31	157.46	15.87	1296	1247	11662	海　南
11100	300	589.75	194.40	386.19	9.16	4801	3729	5008	重　庆
25336	933	1136.49	370.54	749.30	16.65	7527	6421	12025	四　川
11100	1160	350.77	136.65	197.34	16.79	2445	2328	5485	贵　州
10950	100	487.45	196.89	278.93	11.63	5155	4968	5254	云　南
700		62.11	42.43	19.68		778	156	1425	西　藏
6911	100	549.64	300.16	247.01	2.47	6286	6019	5264	陕　西
3208	710	272.55	135.97	121.38	15.21	2115	1849	5624	甘　肃
	170	115.37	97.65		17.71	741	652	951	青　海
2745	560	126.80	40.00	71.66	15.14	898	873	2137	宁　夏
2050	350	332.62	291.17	36.43	5.02	2088	1650	7142	新　疆
800		28.18	16.79	11.39		287	63	527	新疆生产 建设兵团

五、绿色生态数据
Data by Green Ecology

1-14-1　全国历年城市园林绿化情况
National Urban Landscaping in Past Years

计量单位：公顷　　　　　　　Measurement Unit: Hectare

年 份 Year	建成区绿化覆盖面积 Built District Green Coverage Area	建 成 区绿地面积 Built District Area of Green Space	公园绿地面 积 Area of Public Recreational Green Space	公园面积 Park Area	人均公园绿地面积（平方米） Public Recreational Green Space Per Capita (m^2)	建成区绿化覆盖率（%） Green Coverage Rate of Built District (%)	建成区绿地率（%） Green Space Rate of Built District (%)
1981		110037	21637	14739	1.50		
1982		121433	23619	15769	1.65		
1983		135304	27188	18373	1.71		
1984		146625	29037	20455	1.62		
1985		159291	32766	21896	1.57		
1986		153235	42255	30740	1.84	16.9	
1987		161444	47752	32001	1.90	17.1	
1988		180144	52047	36260	1.76	17.0	
1989		196256	52604	38313	1.69	17.8	
1990	246829		57863	39084	1.78	19.2	
1991	282280		61233	41532	2.07	20.1	
1992	313284		65512	45741	2.13	21.0	
1993	354127		73052	48621	2.16	21.3	
1994	396595		82060	55468	2.29	22.1	
1995	461319		93985	72857	2.49	23.9	
1996	493915	385056	99945	68055	2.76	24.43	19.05
1997	530877	427766	107800	68933	2.93	25.53	20.57
1998	567837	466197	120326	73198	3.22	26.56	21.81
1999	593698	495696	131930	77137	3.51	27.58	23.03
2000	631767	531088	143146	82090	3.69	28.15	23.67

注：1. 自2006年起，"公共绿地"统计为"公园绿地"。
　　2. 自2006年起，"人均公共绿地面积"统计为以城区人口和城区暂住人口合计为分母计算的"人均公园绿地面积"，括号内数据约为与往年同口径数据。
Note: 1.Since 2006,Public Green Space is changed to Public Recreatinal Green Space.
　　2.Since 2006,Public Recreational Green Space Per Capita has been calculated based on denominator which combines both permanent and temporary resiclents in urban areas, and the datas in brackets are the same index but calculated by the method of past years.

续表　continued

计量单位: 公顷　　　　　　　　　　　　　　　　　　　　　　　　　　　　Measurement Unit: Hectare

年 份 Year	建成区绿化 覆盖面积 Built District Green Coverage Area	建 成 区 绿地面积 Built District Area of Green Space	公园绿地 面　积 Area of Public Recreational Green Space	公园面积 Park Area	人均公园 绿地面积 （平方米） Public Recreational Green Space Per Capita (m^2)	建成区绿 化覆盖率 （%） Green Coverage Rate of Built District (%)	建成区 绿地率 （%） Green Space Rate of Built District (%)
2001	681914	582952	163023	90621	4.56	28.38	24.26
2002	772749	670131	188826	100037	5.36	29.75	25.80
2003	881675	771730	219514	113462	6.49	31.15	27.26
2004	962517	842865	252286	133846	7.39	31.66	27.72
2005	1058381	927064	283263	157713	7.89	32.54	28.51
2006	1181762	1040823	309544	208056	8.3(9.3)	35.11	30.92
2007	1251573	1110330	332654	202244	8.98	35.29	31.30
2008	1356467	1208448	359468	218260	9.71	37.37	33.29
2009	1494486	1338133	401584	235825	10.66	38.22	34.17
2010	1612458	1443663	441276	258177	11.18	38.62	34.47
2011	1718924	1545985	482620	285751	11.80	39.22	35.27
2012	1812488	1635240	517815	306245	12.26	39.59	35.72
2013	1907490	1719361	547356	329842	12.64	39.86	35.93
2014	2017348	1819960	582392	367926	13.08	40.22	36.29
2015	2105136	1907862	614090	383805	13.35	40.12	36.36
2016	2204040	1992584	653555	416881	13.70	40.30	36.43
2017	2314378	2099120	688441	444622	14.01	40.91	37.11
2018	2419918	2197122	723740	494228	14.11	41.11	37.34
2019	2522931	2285207	756441	502360	14.36	41.51	37.63
2020	2637533	2398085	797912	538477	14.78	42.06	38.24

1-14-2 城市园林绿化(2020年)

地区名称 Name of Regions		绿化覆盖面积 (公顷) Green Coverage Area (hectare)	建成区 Built District	绿地面积 (公顷) Area of Green Space (hectare)
全 国	**National Total**	**3779512**	**2637533**	**3312245**
北 京	Beijing	97141	97141	92683
天 津	Tianjin	47483	43992	43704
河 北	Hebei	116264	95985	98786
山 西	Shanxi	64933	55600	56625
内 蒙 古	Inner Mongolia	72877	51058	68541
辽 宁	Liaoning	220797	113730	147906
吉 林	Jilin	104385	63254	92571
黑 龙 江	Heilongjiang	80198	67367	71526
上 海	Shanghai	170032	46192	164611
江 苏	Jiangsu	337357	208084	305816
浙 江	Zhejiang	199604	133302	179350
安 徽	Anhui	135493	101232	119533
福 建	Fujian	82235	73563	75283
江 西	Jiangxi	84260	78965	77149
山 东	Shandong	299506	235167	262968
河 南	Henan	138690	127423	122110
湖 北	Hubei	122451	113217	105752
湖 南	Hunan	90846	81335	80964
广 东	Guangdong	568920	282760	525545
广 西	Guangxi	122926	66813	74719
海 南	Hainan	19306	15870	17652
重 庆	Chongqing	79620	67405	70680
四 川	Sichuan	148547	135545	130514
贵 州	Guizhou	74639	45781	56825
云 南	Yunnan	57759	51232	51338
西 藏	Tibet	6656	6410	6290
陕 西	Shaanxi	75602	55987	65894
甘 肃	Gansu	35000	32702	30253
青 海	Qinghai	8981	8444	8443
宁 夏	Ningxia	28560	20735	26934
新 疆	Xinjiang	77491	52556	70829
新疆生产 建设兵团	Xinjiang Production and Construction Corps	10951	8687	10450

注：本表中北京市的各项绿化数据均为该市调查面积内数据。

Urban Landscaping(2020)

建成区 Built District	公园绿地 面　积 （公顷） Area of Public Recreational Green Space (hectare)	公园个数 （个） Number of Parks (unit)	门票免费 Free Parks	公园面积 （公顷） Park Area (hectare)	地区名称 Name of Regions
2398085	**797912**	**19823**	**19047**	**538477**	**全　　国**
92683	35720	360	332	35720	北　京
40326	12114	140	133	2858	天　津
87793	29830	779	754	22836	河　北
50600	16375	284	275	12746	山　西
47190	17704	340	333	14325	内　蒙　古
106902	30267	606	588	21061	辽　宁
55875	15736	377	372	12730	吉　林
60567	18085	422	412	11857	黑　龙　江
44307	21981	386	372	3359	上　海
192000	54270	1194	1105	33106	江　苏
120188	38499	1547	1505	22109	浙　江
92762	26520	588	570	18795	安　徽
67473	20763	716	704	16129	福　建
72784	19626	714	705	14856	江　西
213231	70508	1299	1252	47635	山　东
110898	38664	538	507	18463	河　南
100773	31578	520	509	18476	湖　北
72880	21368	456	441	14244	湖　南
256111	114965	4330	4124	84736	广　东
58134	16332	346	332	15222	广　西
14383	4090	128	114	2469	海　南
62036	26571	514	500	16042	重　庆
119302	40440	786	740	24519	四　川
43412	14265	256	246	12933	贵　州
46306	12607	948	897	10280	云　南
6050	1204	155	155	907	西　藏
50899	16554	366	364	10177	陕　西
29281	9825	203	186	6066	甘　肃
7928	2685	64	62	1767	青　海
19596	6304	112	111	3669	宁　夏
47286	10347	182	180	6673	新　疆
8128	2114	167	167	1715	新疆生产 建设兵团

Note: All the greening-related data for Beijing Municipality in the Table are those for the areas surveyed in the city.

县城部分

Statistics for County Seats

一．综合数据
General Data

2-1-1　全国历年县城市政公用设施水平
Level of Service Facilities of National County Seat in Past Years

年 份 Year	供 水 普及率 (%) Water Coverage Rate (%)	燃 气 普及率 (%) Gas Coverage Rate (%)	每万人拥 有公共交 通车辆 (标台) Motor Vehicle for Public Fransport Per 10,000 Persons (standardunit)	人均道路 面 积 (平方米) Road Surface Area Per Capita (m²)	污 水 处理率 (%) Wastewater Treatment Rate (%)	园 林 绿 化			每 万 人 拥有公厕 (座) Number of Public Lavatories per 10,000 Persons (unit)
						人均公园 绿地面积 (平方米) Public Recreational Green Space Per Capita (m²)	建成区绿化 覆 盖 率 (%) Green Coverage Rate of Built District (%)	建成区 绿地率 (%) Green Space Rate of Built District (%)	
2000	84.83	54.41	2.64	11.20	7.55	5.71	10.86	6.51	2.21
2001	76.45	44.55	1.89	8.51	8.24	3.88	13.24	9.08	3.54
2002	80.53	49.69	2.51	9.37	11.02	4.32	14.12	9.78	3.53
2003	81.57	53.28	2.52	9.82	9.88	4.83	15.27	10.79	3.59
2004	82.26	56.87	2.77	10.30	11.23	5.29	16.42	11.65	3.54
2005	83.18	57.80	2.86	10.80	14.23	5.67	16.99	12.26	3.46
2006	76.43	52.45	2.59	10.30	13.63	4.98	18.70	14.01	2.91
2007	81.15	57.33	3.07	10.70	23.38	5.63	20.20	15.41	2.90
2008	81.59	59.11	3.04	11.21	31.58	6.12	21.52	16.90	2.90
2009	83.72	61.66		11.95	41.64	6.89	23.48	18.37	2.96
2010	85.14	64.89		12.68	60.12	7.70	24.89	19.92	2.94
2011	86.09	66.52		13.42	70.41	8.46	26.81	22.19	2.80
2012	86.94	68.50		14.09	75.24	8.99	27.74	23.32	2.09
2013	88.14	70.91		14.86	78.47	9.47	29.06	24.76	2.77
2014	88.89	73.24		15.39	82.12	9.91	29.80	25.88	2.76
2015	89.96	75.90		15.98	85.22	10.47	30.78	27.05	2.78
2016	90.50	78.19		16.41	87.38	11.05	32.53	28.74	2.82
2017	92.87	81.35		17.18	90.21	11.86	34.60	30.74	2.93
2018	93.80	83.85		17.73	91.16	12.21	35.17	31.21	3.13
2019	95.06	86.47		18.29	93.55	13.10	36.64	32.54	3.28
2020	96.66	89.07		18.92	95.05	13.44	37.58	33.55	3.51

注：1.自2006年起，人均和普及率指标按县城人口和县城暂住人口合计为分母计算，以公安部门的户籍统计和暂住人口统
　　计为准。
　　2."人均公园绿地面积"指标2005年及以前年份为"人均公共绿地面积"。
　　3.从2009年起，县城公共交通内容不再统计。

Note: 1.Since 2006,figure in terms of per capita and coverage rate have been calculated based on denominater which combines both
　　　permanent and temporary residents in county seat areas. And the population should come from statistics of police.
　　2.Since 2005, Public Green Space Per Capita is changed to be Public Recreational Green Space Per Capita.
　　3.Since 2009,statistics on county seat public transport have been removed.

2-1-2　全国县城市政公用设施水平(2020年)

地区名称 Name of Regions	人口密度 (人/ 平方公里) Population Density (person/ square kilometer)	人均日生 活用水量 (升) Daily Water Consumption Per Capita (liter)	供　水 普及率 (%) Water Coverage Rate (%)	公共供水 普及率 Public Water Coverage Rate	燃　气 普及率 (%) Gas Coverage Rate (%)	建成区供水 管道密度 (公里/平方公里) Density of Water Supply Pipelines in Built District (kilometer/square kilometer)	人均道路 面　积 (平方米) Road Surface Area Per Capita (m²)	建成区 路网密度 (公里/ 平方公里) Road in Built District (kilometer/ square kilometer)
全　国　National Total	**2107**	**128.53**	**96.66**	**94.69**	**89.07**	**11.56**	**18.92**	**6.88**
河　北　Hebei	2759	110.86	100.00	98.80	98.81	11.45	24.52	8.72
山　西　Shanxi	3418	102.83	96.71	91.26	84.65	11.70	16.68	7.42
内　蒙　古　Inner Mongolia	929	96.16	98.23	97.01	90.12	11.68	29.78	6.87
辽　宁　Liaoning	1508	112.70	97.29	95.86	85.91	11.76	15.15	4.38
吉　林　Jilin	2562	95.60	92.22	88.72	86.56	12.60	14.42	6.10
黑　龙　江　Heilongjiang	2958	92.15	92.69	92.55	60.28	10.15	13.47	6.71
江　苏　Jiangsu	2022	158.93	99.98	99.79	99.95	13.94	21.46	6.60
浙　江　Zhejiang	894	204.30	100.00	100.00	100.00	21.68	24.24	9.13
安　徽　Anhui	1819	136.79	97.14	93.71	95.54	12.42	23.71	6.30
福　建　Fujian	2535	177.76	99.24	99.02	98.03	16.39	18.45	8.78
江　西　Jiangxi	4898	136.25	96.88	96.61	94.14	14.90	20.76	7.88
山　东　Shandong	1373	110.87	99.06	92.38	97.68	7.64	21.98	6.17
河　南　Henan	2646	112.10	93.01	88.51	85.43	7.52	18.50	5.79
湖　北　Hubei	3087	143.96	96.28	95.59	94.11	11.38	18.39	6.70
湖　南　Hunan	3857	148.37	96.70	95.72	90.32	15.02	14.36	6.63
广　东　Guangdong	1519	173.94	92.70	92.49	96.29	14.55	13.36	6.35
广　西　Guangxi	2579	166.91	99.49	97.72	98.79	13.23	21.00	8.40
海　南　Hainan	2816	360.43	95.87	95.81	91.99	8.34	25.04	4.87
重　庆　Chongqing	2725	120.03	97.88	97.88	97.06	12.69	10.21	6.60
四　川　Sichuan	1130	123.14	94.87	94.29	88.36	10.62	13.64	6.03
贵　州　Guizhou	2231	105.67	95.39	95.01	80.17	7.88	18.84	7.49
云　南　Yunnan	3504	116.64	96.85	95.35	55.03	13.57	16.49	7.45
西　藏　Tibet	2723	146.94	87.05	77.98	60.77	8.59	16.28	4.91
陕　西　Shaanxi	3760	97.61	95.50	91.27	88.45	7.49	15.36	6.94
甘　肃　Gansu	4824	80.53	95.14	94.90	76.24	9.38	14.08	5.88
青　海　Qinghai	1908	87.89	97.15	97.15	62.76	9.92	22.42	7.57
宁　夏　Ningxia	3622	112.41	98.99	98.99	75.97	8.66	25.33	6.99
新　疆　Xinjiang	3161	146.23	98.37	98.05	94.63	10.81	23.05	6.57

Level of National County Seat Service Facilities(2020)

建成区道路面积率(%) Road Surface Area Rate of Built District (%)	建成区排水管道密度(公里/平方公里) Density of Sewers in Built District (kilometer/square kilometer)	污水处理率(%) Wastewater Treatment Rate (%)	污水处理厂集中处理率 Centralized Treatment Rate of Wastewater Treatment Plants	人均公园绿地面积(平方米) Public Recreational Green Space Per Capita (m²)	建成区绿化覆盖率(%) Green Coverage Rate of Built District (%)	建成区绿地率(%) Green Space Rate of Built District (%)	生活垃圾处理率(%) Domestic Garbage Treatment Rate (%)	生活垃圾无害化处理率 Domestic Garbage Harmless Treatment Rate	地区名称 Name of Regions
13.15	**9.60**	**95.05**	**94.42**	**13.44**	**37.58**	**33.55**	**99.31**	**98.26**	全　国
17.58	9.64	98.40	98.40	13.76	41.62	37.62	99.96	97.64	河　北
14.13	10.52	96.50	96.28	11.57	40.82	36.12	99.94	99.94	山　西
14.55	8.47	96.23	96.23	21.03	36.82	34.27	99.49	97.50	内 蒙 古
7.36	5.93	95.97	95.89	11.27	19.23	14.58	100.00	90.44	辽　宁
10.77	9.56	95.09	95.09	12.49	34.46	30.40	100.00	100.00	吉　林
7.87	6.12	94.48	94.48	13.08	27.37	24.19	98.71	98.71	黑 龙 江
13.40	12.41	91.40	90.88	14.54	42.64	39.78	100.00	100.00	江　苏
15.81	15.94	97.43	97.29	15.50	43.24	39.04	100.00	100.00	浙　江
14.27	11.40	96.18	96.03	14.57	37.67	33.72	100.00	100.00	安　徽
14.36	13.40	95.06	95.06	15.26	43.39	39.65	100.00	100.00	福　建
14.81	11.56	91.55	90.15	15.54	42.98	38.73	100.00	100.00	江　西
13.24	9.74	97.75	97.75	16.26	40.05	35.80	100.00	100.00	山　东
13.33	9.25	97.01	97.01	11.69	35.78	31.22	99.21	94.77	河　南
13.83	8.83	93.53	93.53	12.25	37.69	32.77	100.00	100.00	湖　北
12.03	9.01	96.57	95.92	11.15	37.72	33.62	99.43	99.43	湖　南
9.49	6.01	91.82	91.80	14.47	35.61	31.81	99.74	99.74	广　东
15.51	11.42	96.40	88.55	13.11	38.56	34.18	100.00	100.00	广　西
10.13	3.76	89.50	89.46	6.82	42.25	29.80	100.00	87.33	海　南
11.95	15.32	99.08	99.08	13.58	41.61	37.73	99.78	99.00	重　庆
11.68	8.92	91.39	88.88	13.71	37.85	34.04	99.77	99.77	四　川
12.68	5.34	91.27	91.21	13.39	36.67	33.80	90.25	90.25	贵　州
13.82	12.99	94.92	94.86	10.79	38.87	34.30	99.88	99.24	云　南
5.86	6.68	31.43	31.43	1.35	6.59	4.01	97.72	92.51	西　藏
12.06	7.96	92.80	92.79	10.93	36.27	31.90	98.68	97.08	陕　西
10.09	9.00	94.41	94.41	11.07	26.76	23.29	99.66	99.66	甘　肃
13.10	9.34	90.72	90.72	7.84	28.01	23.73	95.32	91.57	青　海
14.77	8.46	98.06	98.06	17.38	39.44	36.18	99.58	99.58	宁　夏
11.49	7.05	95.87	95.87	15.12	39.15	35.11	99.26	97.70	新　疆

2-2-1 全国历年县城数量及人口、面积情况

面积计量单位: 平方公里
人口计量单位: 万人

年 份 Year	县个数 Number of Counties	县城人口 County Seat Population	县城暂住人口 County Seat Temporary Population
2000	1674	14157	
2001	1660	9012	
2002	1649	8874	
2003	1642	9235	
2004	1636	9641	
2005	1636	10030	
2006	1635	10963	934
2007	1635	11581	1011
2008	1635	11947	1079
2009	1636	12259	1120
2010	1633	12637	1236
2011	1627	12946	1393
2012	1624	13406	1514
2013	1613	13701	1566
2014	1596	14038	1615
2015	1568	14017	1598
2016	1537	13858	1583
2017	1526	13923	1701
2018	1519	13973	1722
2019	1516	14111	1755
2020	1495	14055	1791

National Changes in Number of Counties, Population and Area in Past Years

Area Measurement Unit: Square Kilometer
Population Measurement Unit: 10,000 persons

县城面积 County Seat Area	建成区面积 Area of Built District	城市建设 用地面积 Area of Urban Construction Land	年　份 Year
53197	13135	8788	2000
57651	10427	9157	2001
56138	10496	9455	2002
53197	11115	10180	2003
53649	11774	11106	2004
63382	12383	12383	2005
76508	13229	13456	2006
93887	14260	14680	2007
130813	14776	15534	2008
154603	15558	15671	2009
175926	16585	16405	2010
93567	17376	16151	2011
94834	18740	17437	2012
86225	19503	17935	2013
79946	20111	18694	2014
75204	20043	18718	2015
72591	19467	18242	2016
71583	19854	18864	2017
70357	20238	19071	2018
76044	20672	19427	2019
75197	20867	19632	2020

2-2-2 全国县城人口和建设用地(2020年)

面积计量单位: 平方公里
人口计量单位: 万人

地区名称 Name of Regions	县面积 County Area	县人口 County Permanent Population	县暂住人口 County Temporary Population	县城面积 County Seat Area	县城人口 County Seat Permanent Population	县城暂住人口 County Seat Temparory Population	建成区面积 Area of Built District	小计 Subtotal
全 国 National Total	7327141	64126	3056	75196.79	14055.43	1790.96	20867.09	19632.14
河 北 Hebei	136230	4032	158	3790.00	943.32	102.19	1405.06	1356.57
山 西 Shanxi	125124	2011	86	1847.09	580.62	50.74	708.28	652.49
内 蒙 古 Inner Mongolia	1049228	1531	127	5482.35	435.83	73.39	993.83	913.15
辽 宁 Liaoning	74824	1020	29	1439.73	205.84	11.28	379.88	362.18
吉 林 Jilin	85748	740	20	699.80	167.50	11.76	233.44	233.20
黑 龙 江 Heilongjiang	217773	1357	37	1260.68	349.87	22.98	599.85	545.73
江 苏 Jiangsu	33080	1992	66	2592.68	484.82	39.36	700.86	681.47
浙 江 Zhejiang	49732	1435	299	5022.22	343.43	105.67	612.57	631.15
安 徽 Anhui	92519	4286	160	4874.69	784.77	102.17	1288.38	1265.72
福 建 Fujian	78389	1784	155	1914.51	414.83	70.42	563.02	559.15
江 西 Jiangxi	121081	2851	88	1722.86	779.67	64.26	1053.73	1020.31
山 东 Shandong	65133	3350		7568.30	979.85	59.60	1560.72	1466.21
河 南 Henan	119053	7164	234	5653.45	1372.09	123.81	1872.48	1759.41
湖 北 Hubei	92363	1973	81	1473.54	414.92	39.96	571.06	533.24
湖 南 Hunan	158255	4523	304	3160.61	1011.10	207.90	1375.46	1260.10
广 东 Guangdong	79301	2240	72	3355.21	467.38	42.12	631.87	581.13
广 西 Guangxi	161657	3205	77	2070.40	486.46	47.53	703.08	658.12
海 南 Hainan	16397	325	27	264.23	57.89	16.52	153.87	131.46
重 庆 Chongqing	39137	933	104	781.74	166.33	46.72	181.86	157.47
四 川 Sichuan	404534	4784	282	10610.79	1019.29	179.89	1326.90	1215.92
贵 州 Guizhou	140665	3089	119	3120.54	625.15	70.90	892.01	791.42
云 南 Yunnan	300133	3256	160	1843.38	561.92	84.09	725.35	690.75
西 藏 Tibet	1198254	274	42	302.53	57.81	24.56	163.36	138.35
陕 西 Shaanxi	150193	2065	82	1395.96	467.26	57.66	624.87	586.95
甘 肃 Gansu	368918	1889	90	837.87	358.66	45.50	491.36	458.58
青 海 Qinghai	481835	354	28	585.83	95.94	15.86	185.09	163.41
宁 夏 Ningxia	38395	337	21	318.11	97.89	17.34	187.92	183.05
新 疆 Xinjiang	1449188	1326	107	1207.69	324.99	56.78	680.93	635.45

注: 本表中山东省人口为常住人口。

County Seat Population and Construction Land(2020)

Area Measurement Unit: Square Kilometer

Population Measurement Unit: 10,000persons

居住用地 Residential	公共管理与公共服务用地 Administration and Public Services	商业服务业设施用地 Commercial and Business Facilities	工业用地 Industrial, Manufacturing	物流仓储用地 Logistics and Warehouse	道路交通设施用地 Road, Street and Transportation	公用设施用地 Municipal Utilities	绿地与广场用地 Green Space and Square	本年征用土地面积 Area of Land Requisition This Year	耕地 Arable Land	地区名称 Name of Regions
6450.08	1718.64	1283.72	2543.97	524.37	3088.92	761.65	3260.79	949.90	428.08	全 国
430.94	92.96	79.57	99.04	21.18	264.51	28.82	339.55	72.12	47.80	河 北
241.89	56.30	34.32	46.36	11.86	110.22	21.17	130.37	5.54	2.49	山 西
288.41	93.86	63.96	80.87	17.70	165.75	34.16	168.44	32.42	2.82	内 蒙 古
153.42	22.05	21.92	65.38	8.14	41.82	21.84	27.61	4.39	3.10	辽 宁
86.86	16.01	13.78	30.77	10.62	31.69	8.28	35.19	6.45	3.32	吉 林
236.43	45.99	27.02	82.02	22.61	64.01	16.46	51.19	7.31	4.08	黑 龙 江
212.59	50.31	41.61	132.54	10.59	103.08	17.38	113.37	22.73	12.34	江 苏
195.13	54.22	41.33	143.48	9.99	84.30	24.08	78.62	90.47	17.71	浙 江
371.48	98.06	84.11	221.62	42.73	222.24	44.46	181.02	63.26	25.79	安 徽
187.25	45.83	36.81	74.26	10.14	91.98	18.66	94.22	26.41	5.79	福 建
301.94	91.48	63.78	158.95	25.12	190.69	34.21	154.14	68.51	23.85	江 西
484.91	116.11	101.06	287.79	38.57	190.77	45.62	201.38	53.03	29.85	山 东
551.04	149.84	106.71	203.41	45.99	303.43	71.65	327.34	56.13	29.32	河 南
166.30	52.69	37.13	77.03	12.16	87.24	21.14	79.55	45.51	22.84	湖 北
413.88	126.48	101.63	165.29	62.54	156.21	66.39	167.68	51.42	14.53	湖 南
208.28	51.89	42.88	92.20	12.93	71.10	25.64	76.21	30.22	7.45	广 东
205.26	52.92	33.17	78.08	16.30	116.54	19.81	136.04	40.79	15.81	广 西
35.78	10.49	7.29	27.52	4.34	25.16	2.34	18.54	8.34	2.31	海 南
55.80	12.28	8.16	16.56	2.29	24.81	6.24	31.33	15.90	10.61	重 庆
402.35	112.43	74.40	179.86	33.64	172.81	54.57	185.86	63.04	37.33	四 川
286.74	78.43	62.72	78.04	24.44	111.15	43.31	106.59	33.85	16.60	贵 州
217.42	72.07	49.68	48.61	16.65	121.82	30.35	134.15	102.41	73.21	云 南
44.67	20.38	11.89	7.60	6.00	25.34	11.36	11.11	6.38	2.10	西 藏
180.54	48.14	37.94	40.45	12.91	91.48	28.53	146.96	16.59	5.91	陕 西
163.78	49.39	31.09	30.98	12.72	66.47	24.60	79.55	12.89	6.27	甘 肃
50.81	20.23	10.74	15.23	4.21	26.04	10.00	26.15	2.53	1.10	青 海
57.32	17.58	16.74	14.04	5.25	31.19	6.62	34.31	7.76	2.69	宁 夏
218.86	60.22	42.28	45.99	22.75	97.07	23.96	124.32	3.50	1.06	新 疆

Note:The Urban District Population of Shandong Province in this table refers to the data of the resident population.

2-3-1　全国历年县城维护建设资金收入

计量单位: 万元

年 份 Year	合 计 Total	城市维护 建 设 税 Urban Maintenance and Construction Tax	城市公用 事业附加 Extra- Charges for Municipal Utilities	中央财政 拨 款 Financial Allocation From Central Government Budget	地方财政 拨 款 Financial Allocation From Local Government Budget
2000	1475288	218874	45849	83404	209840
2001	2339413	284380	47428	68460	378484
2002	3165232	319198	49195	163717	552206
2003	4703331	418992	52225	249529	829818
2004	5178923	482254	58156	159856	1080049
2005	6043627	562226	66971	173570	1405247
2006	5280388	624690	87843	281139	121727
2007	7218834	827944	115565	255958	2022760
2008	9736493	1109612	217957	647527	2927932
2009	13820241	1272267	165937	1125862	5319207
2010	20582038	1520894	215283	1410514	5465372
2011	26240827	1989632	393902	1878689	7259456
2012	31833379	2529027	403033	2937679	9382874
2013	36096026	2840257	523755	2811897	
2014	42036116	2984908	488213	2469729	
2015	36400093	2818465	479671	2199425	
2016	36538509	2940675	473941	2177512	

注: 自2006年起, 县城维护建设资金收入仅包含财政性资金, 不含社会融资。地方财政拨款中包括省、市财政专项拨款和
市级以下财政资金; 其他收入中包括市政公用设施配套费、市政公用设施有偿使用费、土地出让转让金、资产置换收
入及其他财政性资金。

National Revenue of County Seat Maintenance and Construction Fund in Past Years

Measurement Unit: 10,000 RMB

水资源费 Water Resource Fee	国内贷款 Domestic Loan	利用外资 Foreign Investment	企事业单位自筹资金 Self-Raised Funds by Enterprises and Institutions	其他收入 Other Revenues	年 份 Year
5421	177939	52215	243841	437904	2000
7807	261940	65162	535359	690393	2001
9087	374804	76792	775543	844690	2002
14283	603166	131672	1261785	1141861	2003
13112	645288	134733	1203808	1401667	2004
16618	698906	158159	1440598	1521332	2005
30201				4134788	2006
55081				3941526	2007
59435				4774030	2008
88990				5847978	2009
116924				11853051	2010
267072				14452076	2011
283277				16297489	2012
					2013
					2014
					2015
					2016

Note: Since 2006, national revenue of county seats'maintenance and construction fund includes the fund fiscal budget, not including social funds. Local Financial Allocation include province and city special financial allocation, and Other Revenues include fees for expansion of municipal utilites capacity, fees for use of municipal utilities,land transfer income, asset replacement income, and other financial funds.

2-3-2 全国历年县城维护建设资金支出

计量单位：万元

年 份 Year	支出合计 Total	按用途分 By Purpose			供 水 Water Supply	燃 气 Gas Supply	集中供热 Central Heating
		固定资产 投资支出 Expenditure from Investment in Fixed Assets	维护支出 Maintenance Expenditure	其他支出 Other Expenditures			
2000	1543701	1224032	280838	38831	159991	17394	44294
2001	2336929	1746365	333699	256865	203748	42053	81911
2002	3196841	2529827	389298	277716	237120	82809	100698
2003	4661734	3907547	459527	294660	350503	120290	176962
2004	5077874	4221092	559888	296894	418492	132216	226354
2005	5927238	4831130	659689	436419	486596	187305	270791
2006	4971017	3660147	858483	452387	335755	84706	138289
2007	6806308	4796581	1207586	802141	356545	91517	227936
2008	9192126	6722788	1434309	1020577	429065	161977	322488
2009	13182941	10266437	1773499	1322453	581938	195643	471494
2010	20451108	16006075	2348570	2096535	764970	347357	607139
2011	24246084	18101051	3240910	2904123			
2012	33299318	23820534	3899428	5579356			
2013	29418982						
2014	29798372						
2015	27900620						
2016	28732749						

注：从2009年起，县城公共交通内容不再统计。

National Expenditure of County Seat Maintenance and Construction Fund in Past Years

Measurement Unit: 10,000 RMB

按行业分　By Indurstry							
公共交通	道路桥梁	排　水	防　洪	园林绿化	市容环境卫生	其　他	年　份
Public Transportation	Road and Bridge	Sewerage	Flood Control	Landscaping	Environmental Sanitation	Other	Year
21325		113500		81017	69388	226679	2000
56570	900589	170371	84847	167091	105395	524354	2001
69067	1267934	256982	107759	213189	129337	731946	2002
92844	2040028	401354	149805	293000	208242	828706	2003
110896	2068361	447482	164216	386914	206046	916897	2004
136966	2390132	545025	192193	419119	239892	1059219	2005
93385	2192650	533822	144505	434397	284086	729422	2006
164390	2897997	781286	191378	605518	398862	1090879	2007
232359	3904437	1098048	183828	964899	524292	1370733	2008
	5120827	2278356	299657	1375296	1009819	1849911	2009
	8356614	2383811	418408	2543138	1380225	7444422	2010
							2011
							2012
							2013
							2014
							2015
							2016

Note: Since 2009,statistics on county seat public transport have been removed.

2-4-1　按行业分全国历年县城市政公用设施建设固定资产投资

计量单位: 亿元

年 份 Year	本 年 固定 资 产投资总额 Completed Investment of This Year	供　水 Water Supply	燃　气 Gas Supply	集中供热 Central Heating	公共交通 Public Transport	轨道交通 Rail Transit System	道路桥梁 Road and Bridge
2001	337.4	23.6	6.2	8.3	10.0		117.2
2002	412.2	28.1	10.5	13.2	10.6		152.9
2003	555.7	38.8	13.9	18.5	11.8		228.3
2004	656.8	44.4	15.1	24.3	12.3		246.7
2005	719.1	53.4	21.9	29.8	14.1		285.1
2006	730.5	44.3	24.1	28.9	10.3		319.1
2007	812.0	42.3	26.9	42.4	17.7		358.1
2008	1146.1	48.2	35.7	58.5	18.3		532.1
2009	1681.4	78.2	37.0	72.9			690.3
2010	2569.8	88.1	67.2	124.2			1132.1
2011	2859.6	127.6	112.7	155.7			1393.4
2012	3984.7	146.6	137.3	167.8			1934.7
2013	3833.6	164.9	182.3	223.4			1923.5
2014	3572.9	172.6	158.1	187.6			1908.4
2015	3099.8	156.4	112.6	171.0			1663.9
2016	3394.5	160.7	123.1	180.7			1805.8
2017	3634.2	226.3	121.0	194.1			1603.3
2018	3026.0	144.1	103.5	158.6		16.3	1185.0
2019	3076.7	168.1	136.2	133.8		11.1	1312.2
2020	3884.3	232.2	79.7	129.8		14.6	1399.6

注: 1.自2009年开始, 全国县城市政公用设施建设固定资产投资中不再包括公共交通固定资产投资。
　　2.自2013年开始, 全国县城市政公用设施建设固定资产投资中不再包括防洪固定资产投资。

National Fixed Assets Investment of County Seat Service Facilities by Industry in Past Years

Measurement Unit: 100 million RMB

排　水 Sewerage	污水处理 及 其 再 生 利 用 Wastewater Treatment and reused	防　洪 Flood Control	园林绿化 Land- scaping	市容环境 卫　　生 Environmental Sanitation	垃圾处理 Garbage Treatment	地下综合 管　　廊 Utility Tunnel	其　他 Other	年　份 Year
20.4	5.3	11.8	18.2	6.9	1.6		114.9	2001
33.0	12.9	13.7	22.0	10.6	3.9		117.7	2002
44.6	16.1	17.5	30.5	14.9	7.0		136.6	2003
52.5	17.2	19.1	41.0	14.7	4.5		186.7	2004
63.5	24.6	20.4	45.0	17.0	5.4		169.8	2005
72.1	37.0	11.6	46.2	42.2	7.6		132.4	2006
107.1	67.2	17.5	76.0	29.2	16.4		94.9	2007
141.2	80.8	26.4	174.1	37.2	23.2		74.4	2008
305.7	225.7	42.1	222.8	94.7	62.9		137.6	2009
271.1	165.7	45.8	373.6	121.9	83.4		345.9	2010
201.6	108.8	49.3	445.7	172.2	67.2		201.5	2011
229.6	105.0	69.8	581.4	274.6	136.0		442.8	2012
276.1	114.9		587.4	97.3	44.2		378.7	2013
296.1	139.2		521.0	97.4	36.0		231.9	2014
265.8	113.3		480.8	74.0	31.6		175.3	2015
263.0	114.6		500.7	115.9	52.4	22.9	221.8	2016
383.9	104.7		630.6	114.9	53.5	73.5	286.6	2017
367.7	168.0		558.7	134.6	72.2	45.2	312.3	2018
366.6	176.0		482.5	127.3	83.3	46.9	292.1	2019
560.9	306.2		568.2	267.4	199.5	36.9	594.9	2020

Note: 1. Starting from 2009, the national fixed assets investment in the construction of municipal public utilities facilities no longer include the fixed assets investment in public transport.

2. Starting from 2013, the national fixed assets investment in the construction of municipal public utilities facilities no longer include the fixed assets investment in flood prevention.

2-4-2 按行业分全国县城市政公用设施建设固定资产投资(2020年)

计量单位: 万元

地区名称 Name of Regions		本年完成 投 资 Completed Investment of This Year	供 水 Water Supply	燃 气 Gas Supply	集中供热 Central Heating	轨道交通 Rail Transit System	道路桥梁 Road and Bridge	地下综合 管 廊 Utility Tunnel
全 国	**National Total**	**38842864**	**2322499**	**796908**	**1297883**	**145812**	**13996240**	**369184**
河 北	Hebei	2483008	85841	42468	231005		547252	11174
山 西	Shanxi	974359	51606	17403	143283	5500	313455	5776
内 蒙 古	Inner Mongolia	646329	46917	1732	108756		189495	810
辽 宁	Liaoning	91160	15346	878	44633		17870	184
吉 林	Jilin	155583	16292	498	2930		39182	
黑 龙 江	Heilongjiang	403646	112893	21886	73528		65381	
江 苏	Jiangsu	1129845	62811	34400	6410		390896	
浙 江	Zhejiang	2099753	127792	25716			753336	109921
安 徽	Anhui	2719950	212980	104519	7396		1283571	
福 建	Fujian	1830797	108385	69984		18000	864013	41003
江 西	Jiangxi	3719537	186282	74927		1868	1269117	52341
山 东	Shandong	1764325	60095	43714	167639		499355	13626
河 南	Henan	3469200	290698	84964	147275	2011	1144584	6015
湖 北	Hubei	549917	16927	51225			200847	31
湖 南	Hunan	2091337	148670	47680	1300		1006408	2190
广 东	Guangdong	181319	26272	2253			57225	
广 西	Guangxi	3132199	91756	29165			1080108	
海 南	Hainan	249900	752	10623			201354	3539
重 庆	Chongqing	998753	15803	3811			680552	16065
四 川	Sichuan	2042739	90056	23041	23095	92660	868673	18698
贵 州	Guizhou	3661149	180515	31250		22083	1161415	74586
云 南	Yunnan	2015904	112931	33112		36	721238	7468
西 藏	Tibet	48301	4343			599	18781	
陕 西	Shaanxi	507662	9676	17243	45020	375	188465	900
甘 肃	Gansu	703628	65687	6602	99411		240442	1676
青 海	Qinghai	163584	13527		10763		65675	382
宁 夏	Ningxia	163704	4264	2068	64770		36479	
新 疆	Xinjiang	845278	163385	15746	120670	2680	91072	2800

National Investment in Fixed Assets of County Seat Service Facilities by Industry(2020)

Measurement Unit: 10,000 RMB

排 水 Sewerage	污水处理 Wastewater Treatment	污泥处置 Sludge Disposal	再生水利用 Wastewater Recycled and Reused	园林绿化 Landscaping	市容环境卫生 Environmental Sanitation	垃圾处理 Domestic Garbage Treatment	其 他 Other	本年新增固定资产 Newly Added Fixed Assets of This Year	地区名称 Name of Regions
5609088	2905831	116439	155736	5682400	2673817	1995464	5949034	28379136	全 国
633957	212403	1502	12471	381483	453742	419395	96086	1740600	河 北
170429	142369			88599	46968	18344	131341	650642	山 西
146925	72432	1576	28773	72696	19532	13423	59465	302504	内 蒙 古
5893	3912	94		4151	1905	749	300	43747	辽 宁
56895	21503		2397	7833	16961	14828	14991	118607	吉 林
94204	65368	970	5482	14448	13128	8089	8178	210523	黑 龙 江
271539	180448	6500	11083	175466	134258	121821	54065	898299	江 苏
294567	180753	1040	243	211609	231678	204223	345135	1889860	浙 江
405334	130137	2900	1600	369966	151388	100977	184796	1807790	安 徽
221721	110849	315		150685	132070	78154	224937	1148943	福 建
374599	172393	5498	150	307260	233332	126082	1219811	3310771	江 西
311393	112574	2600	2200	329070	230084	148458	109350	1083499	山 东
518037	289084	16799	21718	997741	206364	152264	71511	2714530	河 南
104547	76604	7371		34705	51872	32921	89762	471308	湖 北
277488	218534	19150		82088	151938	112162	373576	1462640	湖 南
49487	36574			4528	7762	7244	33791	83272	广 东
178683	60097	11667		1655791	53512	44106	43184	2969634	广 西
8299	3246	233		2533	9364	8501	13436	29927	海 南
97074	42058	4050		117770	60695	54038	6983	532685	重 庆
347142	185039	5	7757	306887	127801	101365	144686	1919037	四 川
234820	141529	1150		123665	132941	87044	1699875	1398664	贵 州
203406	115071	8724	200	122234	59662	29621	755816	1600591	云 南
11934	11784	50		50	6178	2282	6416	47538	西 藏
135447	72745	9200	1079	26345	26684	24080	57508	389288	陕 西
196693	120572	6400	7762	51420	18646	11700	23052	711005	甘 肃
10061	3554		3600	2855	12683	5839	47638	144375	青 海
20649	9826	500	1755	17218	8186	7151	10071	100578	宁 夏
227865	114373	8145	47466	23304	74483	60604	123274	598276	新 疆

2-5-1 按资金来源分全国历年县城市政公用设施建设固定资产投资

计量单位: 亿元

年份 Year	本年资金 来源合计 Completed Investment of This Year	上 年 末 结余资金 The Balance of The revious Year	本年资金来源		
			小 计 Subtotal	中央财政 拨 款 Financial Allocation From Central Government Budget	地方财政 拨 款 Financial Allocation From Local Government Budget
2001	306.4	7.6	298.8	13.7	47.4
2002	377.1	4.2	372.9	19.7	74.2
2003	518.1	6.8	511.3	30.2	101.6
2004	619.2	7.3	611.9	20.6	133.4
2005	682.8	9.5	673.3	25.1	170.1
2006	755.2	16.9	738.3	54.8	255.4
2007	833.0	13.1	819.8	34.5	346.7
2008	1126.7	16.3	1110.4	52.5	520.4
2009	1682.9	20.8	1662.1	107.0	662.0
2010	2559.8	34.3	2525.5	325.5	985.8
2011	2872.6	31.7	2840.9	152.4	1523.8
2012	3887.7	51.7	3835.9	205.0	2073.7
2013	3683.0	60.7	3622.2	135.8	1070.7
2014	3690.4	63.2	3627.2	111.8	968.5
2015	3011.2	52.8	2958.4	121.7	930.7
2016	3190.9	30.7	3160.2	120.6	1074.4
2017	3610.8	86.6	3524.2	107.0	1035.3
2018	3222.3	102.6	3119.6	132.0	797.0
2019	3900.5	148.0	3752.5	169.3	892.6
2020	4371.0	210.6	4160.4	252.2	1135.1

注: 1.自2013年起, "本年资金来源合计"为"本年实际到位资金合计"。
    2.自2013年起, "中央财政拨款"为"中央预算资金", "地方财政拨款"为除"中央预算资金"外的"国家预算资金"合计。

National Fixed Assets Investment of County Seat Service Facilities by Capital Source in Past Years

Measurement Unit: 100 million RMB

Sources of Fund					
国内贷款 Domestic Loan	债 券 Securities	利用外资 Foreign Investment	自筹资金 Self-Raised Funds	其他资金 Other Funds	年份 Year
33.2	1.5	23.4	116.1	63.5	2001
46.3	1.2	12.6	149.5	69.4	2002
69.0	1.6	25.1	202.2	81.7	2003
84.1	2.2	38.7	222.1	110.7	2004
76.9	2.2	39.9	247.3	111.9	2005
89.6	1.5	26.2	234.2	76.6	2006
88.1	2.2	26.3	240.3	81.8	2007
107.6	1.4	28.0	297.0	103.5	2008
298.6	11.7	32.9	385.5	164.4	2009
332.6	4.1	44.3	606.6	226.6	2010
278.0	3.2	31.8	644.6	207.2	2011
318.1	2.3	85.0	877.2	274.7	2012
277.2	7.7	56.1	1610.4	464.3	2013
315.5	4.1	23.3	1767.7	436.3	2014
222.2	5.1	33.1	1233.3	412.2	2015
221.6	14.0	19.0	1339.4	371.3	2016
378.9	32.9	21.2	1372.3	576.5	2017
171.5	38.4	16.8	1307.4	656.4	2018
174.6	92.1	32.0	1457.7	934.1	2019
224.0	406.7	29.0	1169.9	943.6	2020

Note: 1.Since 2013, Completed Investment of This Year is changed to The Total Funds Actually Available for The Reported Year.
2.Since 2013, Financial Allocation from Central Government Budget is changed to Central Budgetary Fund, and Financial Allocation from Local Government Budget is State Budgetary Fund excluding Central Budgetary Fund.

2-5-2　按资金来源分全国县城市政公用设施建设固定资产投资(2020年)

计量单位: 万元

地区名称 Name of Regions	本市实际到位资金合计 The Total Funds Actually Available for The Reported Year	上年末结余资金 The Balance of The Previous Year	本年资金来源			
			小　计 Subtotal	国家预算资金 State Budgetary Fund	中央预算资金 Central Budgetary Fund	国内贷款 Domestic Loan
全　国　National Total	**43709750**	**2105568**	**41604181**	**13872697**	**2521622**	**2239594**
河　北　Hebei	2369237	24216	2345021	754380	59718	110626
山　西　Shanxi	910313	48566	861747	210792	26063	104899
内 蒙 古　Inner Mongolia	903321	78532	824790	144995	67190	8241
辽　宁　Liaoning	159369	780	158589	98389	30017	6152
吉　林　Jilin	331760	9387	322373	105665	52730	12112
黑 龙 江　Heilongjiang	646072	22439	623633	207285	142248	1213
江　苏　Jiangsu	958905	9612	949293	259564	439	27000
浙　江　Zhejiang	2296332	26478	2269855	572326	26411	127925
安　徽　Anhui	2716877	73229	2643648	1240388	17500	111384
福　建　Fujian	1765296	33523	1731773	450636	106079	32221
江　西　Jiangxi	5107494	307057	4800437	1590340	123302	301186
山　东　Shandong	1498877	39393	1459484	399955	8815	76123
河　南　Henan	3508328	38964	3469364	1515445	144534	132162
湖　北　Hubei	749793	32844	716950	156497	80240	49640
湖　南　Hunan	2735695	71349	2664346	563765	209051	153570
广　东　Guangdong	270831	45134	225697	48706	1100	10088
广　西　Guangxi	3134388	102043	3032345	2613934	100360	73176
海　南　Hainan	347240	30725	316515	250777	137396	646
重　庆　Chongqing	1076176	2100	1074076	522455	55361	311425
四　川　Sichuan	2395762	166097	2229666	263837	161304	170257
贵　州　Guizhou	4290689	633511	3657178	296088	88612	213708
云　南　Yunnan	2124703	106415	2018289	525570	221404	85589
西　藏　Tibet	126807	17885	108922	73749	33271	
陕　西　Shaanxi	673938	24150	649788	214290	72431	13400
甘　肃　Gansu	848742	63991	784751	191847	157359	81815
青　海　Qinghai	327565	77995	249570	157824	116809	2565
宁　夏　Ningxia	153874	2945	150929	66454	40313	
新　疆　Xinjiang	1281366	16210	1265156	376744	241565	22471

National Investment in Fixed Assets of County Seat Service Facilities by Capital Source(2020)

Measurement Unit: 10,000 RMB

Sources of Fund				各项应付款	地区名称
债 券 Securities	利用外资 Foreign Investment	自筹资金 Self-Raised Funds	其他资金 Other Funds	Sum Payable This Year	Name of Regions
4066593	290351	11699171	9435777	7562679	全　国
503895		792069	184051	359724	河　北
88374	180	336100	121402	271482	山　西
200513		254518	216523	132765	内 蒙 古
13781	350	30742	9175	13941	辽　宁
148405		54011	2180	38666	吉　林
191288		179518	44328	52541	黑 龙 江
59918		400129	202682	212076	江　苏
60402		1057011	452191	178975	浙　江
117492	9333	637478	527574	393001	安　徽
93013		894758	261145	540692	福　建
72661	104677	1313885	1417688	545328	江　西
247783	10504	411256	313862	425373	山　东
311030	14944	752309	743473	356934	河　南
80978		229862	199974	190808	湖　北
108361	62039	1404926	371685	206823	湖　南
40545	3940	57328	65089	58115	广　东
48034	2910	189789	104501	149317	广　西
3840		9005	52246	2156	海　南
122631	2438	114478	650	19713	重　庆
138202	19764	695025	942581	468797	四　川
121281		928867	2097233	602230	贵　州
276478	43538	466429	620684	1747091	云　南
5000	1500	7539	21134	11244	西　藏
49127	3500	213320	156151	155911	陕　西
274415	4697	111967	120009	267002	甘　肃
42006		18509	28667	7711	青　海
61370		10773	12333	37600	宁　夏
585770	6037	127568	146566	116662	新　疆

二、居民生活数据

Data by Residents Living

2-6-1 全国历年县城供水情况

年 份 Year	综 合 生产能力 (万立方米／日) Integrated Production Capacity (10,000 m³/day)	供水管道 长 度 (公里) Length of Water Supply Pipelines (km)	供水总量 (万立方米) Total Quantity of Water Supply (10,000 m³)	生活用量 Residential Use
2000	3662	70046	593588	310331
2001	3754	77316	577948	327081
2002	3705	78315	567915	338689
2003	3400	87359	606189	363311
2004	3680	92867	653814	395181
2005	3862	98980	676548	409045
2006	4207	113553	746892	407389
2007	5744	131541	794495	448943
2008	5976	142507	826300	459866
2009	4775	148578	856291	484644
2010	4683	159905	925705	509266
2011	5174	173452	977115	534918
2012	5446	186500	1020298	565835
2013	5239	194465	1038662	584759
2014	5437	203514	1063270	600436
2015	5769	214736	1069191	612383
2016	5421	211355	1064966	609171
2017	6443	234466	1128373	636403
2018	7415	242537	1145082	660469
2019	6304	258602	1190883	697214
2020	6451	272990	1190206	718585

注: 自2006年起, 供水普及率指标按城区人口和城区暂住人口合计为分母计算。

National County Seat Water Supply in Past Years

用水人口 （万人） Population with Access to Water Supply (10,000 persons)	人　均　日 生活用水量 （升） Daily Water Consumption Per Capita (liter)	供　水 普及率 (%) Water Coverage Rate (%)	年　份 Year
6931.2	122.7	84.83	2000
6889.2	130.1	76.45	2001
7145.7	129.9	80.53	2002
7532.9	132.1	81.57	2003
7931.1	136.5	82.26	2004
8342.2	134.3	83.18	2005
9093.4	122.7	76.43	2006
10218.7	120.4	81.15	2007
10628.4	119.4	81.59	2008
11200.7	118.6	83.72	2009
11811.4	118.9	85.14	2010
12345.0	118.7	86.09	2011
12971.0	119.5	86.94	2012
13456.0	119.1	88.14	2013
13913.4	118.2	88.89	2014
14048.3	119.4	89.96	2015
13974.9	119.4	90.50	2016
14508.9	120.2	92.87	2017
14722.5	122.9	93.80	2018
15082.0	126.7	95.06	2019
15316.7	128.5	96.66	2020

Note: Since 2006, water coverage rate has been calculated based on denominator which combines both permanent and temporary residents in urban areas, and the datas of brackets are the same index but calculated by the method of past years.

2-6-2 县城供水(2020年)

地区名称 Name of Regions		综合 生产能力 (万立方 米/日) Integrated Production Capacity (10,000 m³/ day)	地下水 Underground Water	供水管道 长度 (公里) Length of Water Supply Pipelines (km)	建成区 in Built District	供水总量 (万立方米) Total Quantity of Water Supply (10,000 m³)
全 国	**National Total**	**6450.63**	**1581.14**	**272990.01**	**241198.21**	**1190205.80**
河 北	Hebei	401.85	169.99	17426.58	16088.77	70555.26
山 西	Shanxi	155.26	116.45	9662.23	8286.87	33938.40
内 蒙 古	Inner Mongolia	160.62	156.85	12075.59	11611.68	30708.12
辽 宁	Liaoning	72.12	18.03	5004.55	4468.59	15442.74
吉 林	Jilin	52.92	15.92	2998.44	2941.55	11099.57
黑 龙 江	Heilongjiang	114.23	90.38	6207.47	6088.55	18827.07
江 苏	Jiangsu	292.13	16.01	11877.01	9770.60	51808.07
浙 江	Zhejiang	366.47	0.15	17550.47	13283.58	67050.10
安 徽	Anhui	352.97	86.79	17494.13	16004.17	76016.16
福 建	Fujian	275.38	9.75	11406.94	9230.55	49029.13
江 西	Jiangxi	405.07	51.78	16884.91	15703.67	68223.00
山 东	Shandong	432.30	189.08	12823.32	11923.75	87797.12
河 南	Henan	594.42	276.38	14948.31	14072.57	93790.94
湖 北	Hubei	234.98	0.92	7270.80	6499.42	38597.95
湖 南	Hunan	487.23	29.08	22767.33	20659.32	102422.45
广 东	Guangdong	232.45		10712.76	9195.91	48133.49
广 西	Guangxi	256.65	20.32	9422.38	9301.60	49543.62
海 南	Hainan	171.15	11.73	2308.14	1282.90	14514.58
重 庆	Chongqing	76.55		2314.92	2308.62	14782.38
四 川	Sichuan	358.23	15.89	15196.10	14089.56	78429.05
贵 州	Guizhou	207.53	19.74	10198.48	7024.80	37764.68
云 南	Yunnan	176.95	12.00	11373.83	9839.73	39483.33
西 藏	Tibet	85.37	19.89	1623.17	1402.65	5273.91
陕 西	Shaanxi	157.88	90.81	5520.59	4682.99	27324.44
甘 肃	Gansu	89.35	33.69	5175.17	4608.87	16214.91
青 海	Qinghai	40.12	10.10	1945.76	1836.61	5010.63
宁 夏	Ningxia	44.77	28.83	1986.08	1627.65	8383.64
新 疆	Xinjiang	155.68	90.58	8814.55	7362.68	30041.06

County Seat Water Supply(2020)

生产运营用水 The Quantity of Water for Production and Operation	公共服务用水 The Quantity of Water for Public Service	居民家庭用水 The Quantity of Water for Household Use	其他用水 The Quantity of Water for Other Purposes	用水户数 (户) Number of Households with Access to Water Supply (unit)	家庭用户 Household User	用水人口 (万人) Population with Access to Water Supply (10,000 persons)	地区名称 Name of Regions
256547.18	114575.10	597371.05	53133.22	52608494	45953219	15316.65	全　　国
17543.13	6702.47	35480.32	2188.50	3628333	3310949	1045.51	河　　北
7535.24	4081.93	18586.22	643.76	1593096	1478268	610.57	山　　西
7671.15	3898.04	13276.08	1721.05	2087586	1842480	500.23	内 蒙 古
1527.39	1698.79	6917.08	536.31	1030457	924847	211.23	辽　　宁
1700.86	1107.93	4592.72	806.35	703220	639266	165.32	吉　　林
3203.19	2021.73	9439.36	533.34	1555930	1369573	345.61	黑 龙 江
13151.26	6507.44	23393.25	2206.56	2346482	2138036	524.09	江　　苏
21192.12	5513.41	27951.55	3355.39	2282838	2023722	449.10	浙　　江
21190.71	7333.02	35515.88	2621.69	3179589	2813662	861.59	安　　徽
6303.83	5703.68	25505.26	2173.78	1875180	1645623	481.56	福　　建
13808.57	5966.24	34443.56	3402.88	3154049	2752785	817.59	江　　西
35659.95	7129.49	34450.16	3619.29	2547699	2333855	1029.64	山　　东
22323.74	11742.78	44892.12	5640.86	4240364	3582045	1391.40	河　　南
6479.13	3301.68	19269.84	1464.96	1475692	1278365	437.98	湖　　北
16671.79	7742.41	55074.91	4108.30	4047016	3033055	1178.72	湖　　南
7523.11	3506.95	26421.16	2798.56	1572994	1337476	472.32	广　　东
10001.31	4335.23	27955.51	1016.88	1542885	1433528	531.28	广　　西
2375.93	1227.97	8097.96	1312.71	244984	223884	71.34	海　　南
1715.48	971.78	8164.38	1075.62	862537	760505	208.54	重　　庆
12266.16	7194.68	43286.85	3189.19	4192924	3633006	1137.63	四　　川
4990.39	2054.37	23422.12	928.81	2081420	1730921	663.96	贵　　州
6963.75	3814.66	22361.41	970.60	1830103	1587887	625.68	云　　南
407.63	290.26	2639.28	149.08	188154	172930	71.70	西　　藏
5767.82	3001.30	14761.15	1288.45	1327589	1217961	501.32	陕　　西
2227.19	2073.46	9206.71	868.99	1011281	909011	384.53	甘　　肃
593.47	798.31	2677.80	230.15	296969	253969	108.61	青　　海
1938.73	972.44	3691.25	859.22	455347	373023	114.07	宁　　夏
3814.15	3882.65	15897.16	3421.94	1253776	1152587	375.53	新　　疆

2-6-3　县城供水(公共供水)(2020年)

地区名称 Name of Regions	综合生产能力 (万立方米/日) Integrated Production Capacity (10,000 m³/day)	地下水 Underground Water	水厂个数 (个) Number of Water Plants (unit)	地下水 Underground Water	供水管道长度 (公里) Length of Water Supply Pipelines (km)	供水总量（万立方米） 合计 Total	售水量 小计 Subtotal	生产运营用水 The Quantity of Water for Production and Operation
全　国　National Total	**5529.75**	**1074.82**	**2490**	**888**	**260849.36**	**1067371.26**	**898792.01**	**179413.36**
河　北　Hebei	347.90	124.85	175	97	16280.35	60636.45	51995.61	9730.71
山　西　Shanxi	108.04	72.90	125	98	8181.39	25157.35	22066.10	3603.59
内　蒙　古　Inner Mongolia	124.34	121.36	122	114	11712.20	23549.30	19407.50	2904.53
辽　宁　Liaoning	68.32	14.23	31	14	4849.40	15138.74	10375.57	1376.39
吉　林　Jilin	46.45	10.80	23	10	2887.63	10190.07	7298.36	1208.14
黑　龙　江　Heilongjiang	98.53	74.68	63	54	6023.77	17377.25	13747.80	2282.26
江　苏　Jiangsu	257.70	4.70	25	1	11527.45	47164.15	40614.59	10233.83
浙　江　Zhejiang	344.27		55		17448.45	65227.82	56190.19	19829.84
安　徽　Anhui	304.23	46.73	78	18	16879.31	63789.37	54434.51	11873.53
福　建　Fujian	266.90	9.40	76	9	11350.54	48420.21	39077.63	5889.03
江　西　Jiangxi	378.05	32.00	117	11	16685.96	67470.61	56868.86	13609.57
山　东　Shandong	300.76	90.95	136	70	11303.95	60805.61	53867.38	15484.69
河　南　Henan	407.89	179.78	151	89	13524.60	71010.06	61818.62	9127.66
湖　北　Hubei	227.21		67	2	7205.27	37962.53	29880.19	6234.81
湖　南　Hunan	452.90	22.50	115	10	22044.03	98632.81	79807.77	15508.63
广　东　Guangdong	217.05		56		10515.89	47811.19	39927.48	7478.81
广　西　Guangxi	224.26	6.62	88	14	9308.76	46111.61	39876.92	7380.90
海　南　Hainan	51.84		17		1783.69	9426.76	7926.75	1433.60
重　庆　Chongqing	76.55		33		2314.92	14782.38	11927.26	1715.48
四　川　Sichuan	344.40	9.92	176	11	14996.15	75946.74	63454.57	10635.66
贵　州　Guizhou	207.31	19.60	146	20	10179.28	37735.03	31366.04	4986.67
云　南　Yunnan	168.32	7.33	169	17	11103.95	38480.77	33107.86	6598.55
西　藏　Tibet	72.57	13.41	78	49	1433.37	4645.80	2858.14	326.63
陕　西　Shaanxi	119.14	63.44	122	54	4861.78	22530.05	20024.33	2907.76
甘　肃　Gansu	84.75	30.26	87	35	4987.07	15631.61	13793.05	1973.09
青　海　Qinghai	39.62	9.60	40	11	1938.76	4922.64	4211.74	508.48
宁　夏　Ningxia	40.27	24.33	22	12	1879.23	7261.29	6339.29	1034.92
新　疆　Xinjiang	150.18	85.43	97	68	7642.21	29553.06	26527.90	3535.60

County Seat Water Supply　(Public Water Suppliers)(2020)

| Total Quantity of Water Supply(10,000m³) | | | | | | 用水户数 | 家庭用户 | 用水人口 | 地区名称 |
| Water Sold | | | | | | (户) | | (万人) | |
公共服务用水 The Quantity of Water for Public Service	居民家庭用水 The Quantity of Water for Household Use	其他用水 The Quantity of Other Purposes	免费供水量 The Quantity of Free Water Supply	生活用水 Domestic Water Use	漏损水量 The Lossed Water	Number of Households with Access to Water Supply (unit)	家庭用户 Household User	Population with Access to Water Supply (10,000 persons)	Name of Regions
101960.34	**571012.82**	**46405.49**	**34503.64**	**6638.52**	**134075.61**	**51029575**	**44778680**	**15005.21**	全　　国
6059.78	34107.11	2098.01	1517.95	122.17	7122.89	3549674	3251060	1032.96	河　　北
3208.92	14830.53	423.06	688.49	248.97	2402.76	1413400	1318729	576.16	山　　西
2969.79	12421.87	1111.31	944.07	383.90	3197.73	2049254	1822424	493.99	内 蒙 古
1655.09	6807.78	536.31	494.16	72.80	4269.01	1024219	918797	208.13	辽　　宁
1032.58	4361.15	696.49	526.39	68.00	2365.32	677958	616064	159.04	吉　　林
1689.77	9323.86	451.91	617.69	163.02	3011.76	1538337	1354510	345.06	黑 龙 江
6085.68	22204.32	2090.76	1114.82	501.80	5434.74	2337141	2131374	523.07	江　　苏
5053.41	27951.55	3355.39	1756.52	24.69	7281.11	2282741	2023722	449.10	浙　　江
6646.75	33910.34	2003.89	1805.89	168.91	7548.97	3080258	2720986	831.15	安　　徽
5633.28	25386.46	2168.86	1876.35	36.42	7466.23	1871066	1642605	480.48	福　　建
5945.99	33927.78	3385.52	1942.40	248.79	8659.35	3135253	2734724	815.34	江　　西
5606.50	30882.45	1893.74	555.71	85.67	6382.52	2313177	2176354	960.24	山　　东
7536.42	40645.00	4509.54	1943.91	294.17	7247.53	3996123	3384734	1324.04	河　　南
3146.58	19081.64	1417.16	2277.49	442.60	5804.85	1459207	1263957	434.81	湖　　北
7085.54	53393.40	3820.20	6186.66	1015.33	12638.38	3831854	2941754	1166.78	湖　　南
3463.95	26192.06	2792.66	901.63	57.76	6982.08	1537386	1334573	471.22	广　　东
4291.93	27301.31	902.78	491.47	75.17	5743.22	1514710	1405848	521.83	广　　西
832.15	5242.12	418.88	137.42	59.27	1362.59	172846	159977	71.29	海　　南
971.78	8164.38	1075.62	763.80		2091.32	862537	760505	208.54	重　　庆
7002.74	42677.01	3139.16	2003.03	650.03	10489.14	4168761	3610460	1130.76	四　　川
2049.53	23406.20	923.64	679.32	132.46	5689.67	2078808	1729553	661.31	贵　　州
3709.20	21864.55	935.56	1891.56	461.27	3481.35	1794134	1569287	616.00	云　　南
238.05	2172.52	120.94	1243.49	916.05	544.17	165452	152285	64.23	西　　藏
2459.79	13509.11	1147.67	470.93	97.86	2034.79	1172020	1077693	479.10	陕　　西
1993.98	9100.81	725.17	460.40	23.02	1378.16	1008356	906451	383.56	甘　　肃
798.31	2677.80	227.15	273.17	8.21	437.73	296969	253969	108.61	青　　海
969.94	3691.25	643.18	72.14	16.50	849.86	455108	373023	114.07	宁　　夏
3822.91	15778.46	3390.93	866.78	263.68	2158.38	1242826	1143262	374.34	新　　疆

2-6-4 县城供水(自建设施供水)(2020年)

地区名称 Name of Regions	综 合 生产能力 (万立方米/日) Integrated Production Capacity (10,000 m³/day)	地下水 Underground Water	供水管道 长 度 (公里) Length of Water Supply Pipelines (km)	建成区 in Built District	供水总量(万立方米)	
					合计 Total	生产运营 用 水 The Quantity of Water for Production and Operation
全 国 National Total	920.88	506.32	12140.65	8850.80	122834.54	77133.82
河 北 Hebei	53.95	45.14	1146.23	1030.58	9918.81	7812.42
山 西 Shanxi	47.22	43.55	1480.84	1074.36	8781.05	3931.65
内 蒙 古 Inner Mongolia	36.28	35.49	363.39	299.26	7158.82	4766.62
辽 宁 Liaoning	3.80	3.80	155.15	133.15	304.00	151.00
吉 林 Jilin	6.47	5.12	110.81	100.56	909.50	492.72
黑 龙 江 Heilongjiang	15.70	15.70	183.70	118.60	1449.82	920.93
江 苏 Jiangsu	34.43	11.31	349.56	299.51	4643.92	2917.43
浙 江 Zhejiang	22.20	0.15	102.02	7.02	1822.28	1362.28
安 徽 Anhui	48.74	40.06	614.82	342.22	12226.79	9317.18
福 建 Fujian	8.48	0.35	56.40	3.00	608.92	414.80
江 西 Jiangxi	27.02	19.78	198.95	182.49	752.39	199.00
山 东 Shandong	131.54	98.13	1519.37	1151.87	26991.51	20175.26
河 南 Henan	186.53	96.60	1423.71	1062.74	22780.88	13196.08
湖 北 Hubei	7.77	0.92	65.53	27.00	635.42	244.32
湖 南 Hunan	34.33	6.58	723.30	701.74	3789.64	1163.16
广 东 Guangdong	15.40		196.87	33.00	322.30	44.30
广 西 Guangxi	32.39	13.70	113.62	113.62	3432.01	2620.41
海 南 Hainan	119.31	11.73	524.45	398.95	5087.82	942.33
重 庆 Chongqing						
四 川 Sichuan	13.83	5.97	199.95	121.73	2482.31	1630.50
贵 州 Guizhou	0.22	0.14	19.20	0.48	29.65	3.72
云 南 Yunnan	8.63	4.67	269.88	224.41	1002.56	365.20
西 藏 Tibet	12.80	6.48	189.80	162.11	628.11	81.00
陕 西 Shaanxi	38.74	27.37	658.81	437.60	4794.39	2860.06
甘 肃 Gansu	4.60	3.43	188.10	129.34	583.30	254.10
青 海 Qinghai	0.50	0.50	7.00	7.00	87.99	84.99
宁 夏 Ningxia	4.50	4.50	106.85	91.12	1122.35	903.81
新 疆 Xinjiang	5.50	5.15	1172.34	597.34	488.00	278.55

County Seat Water Supply （Suppliers with Self-Built Facilities)(2020)

Total Quantity of Water Supply(10,000m³)			用水户数 （户） Number of Households with Access to Water Supply (unit)	家庭用户 Household User	用水人口 （万人） Population with Access to Water Supply (10,000 persons)	地区名称 Name of Regions
公共服务 用 水 The Quantity of Water for Public Service	居民家庭 用 水 The Quantity of Water for Household Use	其他用水 The Quantity of Water for Other Purposes				
12614.76	26358.23	6727.73	1578919	1174539	311.44	全　国
642.69	1373.21	90.49	78659	59889	12.55	河　北
873.01	3755.69	220.70	179696	159539	34.41	山　西
928.25	854.21	609.74	38332	20056	6.24	内 蒙 古
43.70	109.30		6238	6050	3.10	辽　宁
75.35	231.57	109.86	25262	23202	6.28	吉　林
331.96	115.50	81.43	17593	15063	0.55	黑 龙 江
421.76	1188.93	115.80	9341	6662	1.02	江　苏
460.00		—	97			浙　江
686.27	1605.54	617.80	99331	92676	30.44	安　徽
70.40	118.80	4.92	4114	3018	1.08	福　建
20.25	515.78	17.36	18796	18061	2.25	江　西
1522.99	3567.71	1725.55	234522	157501	69.40	山　东
4206.36	4247.12	1131.32	244241	197311	67.36	河　南
155.10	188.20	47.80	16485	14408	3.17	湖　北
656.87	1681.51	288.10	215162	91301	11.94	湖　南
43.00	229.10	5.90	35608	2903	1.10	广　东
43.30	654.20	114.10	28175	27680	9.45	广　西
395.82	2855.84	893.83	72138	63907	0.05	海　南
						重　庆
191.94	609.84	50.03	24163	22546	6.87	四　川
4.84	15.92	5.17	2612	1368	2.65	贵　州
105.46	496.86	35.04	35969	18600	9.68	云　南
52.21	466.76	28.14	22702	20645	7.47	西　藏
541.51	1252.04	140.78	155569	140268	22.22	陕　西
79.48	105.90	143.82	2925	2560	0.97	甘　肃
		3.00				青　海
2.50		216.04	239			宁　夏
59.74	118.70	31.01	10950	9325	1.19	新　疆

2-7-1　全国历年县城节约用水情况

年 份 Year	计划用水量 （万立方米） Planned Quantity of Water Use (10,000 m³)	新水取用量 （万立方米） Fresh Water Used (10,000 m³)
2000	115591	109069
2001	161795	133634
2002	110702	91610
2003	120551	94938
2004	123176	98004
2005	155933	118727
2006		122700
2007		127318
2008		122275
2009		139735
2010		157563
2011		119923
2012		180603
2013		136223
2014		140966
2015		126923
2016		115707
2017		106174
2018		111385
2019		140990
2020		188350

注：自2006年起，不统计计划用水量指标。

National County Seat Water Conservation in Past Years

工业用水 重复利用量 （万立方米） Quantity of Industrial Water Recycled (10,000 m³)	节约用水量 （万立方米） Water Saved (10,000 m³)	年 份 Year
95763	20322	2000
24206	28161	2001
26321	19092	2002
24815	25613	2003
25790	25163	2004
32852	37206	2005
32523	21546	2006
99648	22675	2007
53278	20480	2008
130864	23135	2009
133337	35963	2010
95932	26165	2011
52051	30853	2012
159152	22286	2013
164833	30029	2014
173337	32419	2015
165369	23086	2016
167080	24873	2017
189581	26618	2018
169218	28609	2019
104108	48132	2020

Note: From 2006, "Planned Quantity of Water Use" has not been counted.

2-7-2 县城节约用水(2020年)

计量单位: 万立方米

地区名称 Name of Regions	计划用水户数(户) Planned Water Consumers (households)	自备水计划用水户数 Planned Self-produced Water Consumers	计划用水户实际用水量		新水取水量 Fresh Water Used	工 业 Industry	重复利用量 Water Reused
			合 计 Total	工 业 Industry			
全 国 National Total	1506530	63731	310999	165668	188350	61559	122649
河 北 Hebei	27495	37	760	129	717	86	43
山 西 Shanxi	205561	4858	14176	3336	12036	1927	2141
内 蒙 古 Inner Mongolia	80200	5	983	206	704	140	279
辽 宁 Liaoning	10600	61	2191	756	1787	496	404
吉 林 Jilin	58	20	830	99	753	47	77
黑 龙 江 Heilongjiang	45276		336	133	317	114	19
江 苏 Jiangsu	2209	171	35888	26098	15600	7884	20288
浙 江 Zhejiang	2768	216	13080	9077	7671	4420	5409
安 徽 Anhui							
福 建 Fujian	4935	9	4693	1969	3549	1394	1144
江 西 Jiangxi			1034	375	517	200	517
山 东 Shandong	41786	12312	59006	46631	17782	9302	41225
河 南 Henan	80319	20104	24998	10851	19734	7360	5264
湖 北 Hubei	12850	920	15141	1738	13922	1617	1219
湖 南 Hunan	24290	5375	19369	2609	15993	2423	3376
广 东 Guangdong	232341	2047	9668	248	9570	236	98
广 西 Guangxi							
海 南 Hainan	27735	56	13791	10566	13781	10556	10
重 庆 Chongqing	93	60	6569	1911	3257	289	3312
四 川 Sichuan	308691	840	10852	4199	9402	3276	1450
贵 州 Guizhou	239	41	10086	2996	8280	1531	1806
云 南 Yunnan	49033	6550	14857	1151	14136	682	721
西 藏 Tibet	796	796	591		591		
陕 西 Shaanxi	68090	7274	41001	37605	8728	5478	32273
甘 肃 Gansu	79185	1852	2716	682	1867	186	848
青 海 Qinghai							
宁 夏 Ningxia	58	20	3352	1701	3047	1396	305
新 疆 Xinjiang	201922	107	5029	601	4609	520	421

County Seat Water Conservation(2020)

Actual Quantity of Water Used			节约用水量 Water Saved			节水措施投资总额（万元）Total Investment in Water-Saving Measures (10,000 RMB)	地区名称 Name of Regions
工 业 Industry	超计划定额用水量 Water Quantity Consumed in Excess of Quota	重复利用率（%）Reuse Rate（%）	工 业 Industry		工 业 Industry		
104108	3783	39.44	62.84	48132	34557	65797	全 国
43	10	5.72	33.58	43	43	540	河 北
1409	173	15.10	42.22	851	343	15641	山 西
66		28.38	32.04	44	21	715	内 蒙 古
260		18.46	34.44	439	275	551	辽 宁
51		9.29	52.10	140	51	40	吉 林
19		5.65	14.29	16		1000	黑 龙 江
18214		56.53	69.79	14459	12432	17283	江 苏
4657	164	41.35	51.30	1018	529	2280	浙 江
							安 徽
575	56	24.38	29.20	135	118	10	福 建
175		50.02	46.60				江 西
37330	48	69.86	80.05	8663	7885	4308	山 东
3491	904	21.06	32.18	4976	3081	1991	河 南
121		8.05	6.96	932	121	65	湖 北
186		17.43	7.13	3207	82	2591	湖 南
12		1.01	4.84	25	14	31	广 东
							广 西
10		0.07	0.09	2379	2095	28	海 南
1622	2244	50.42	84.86	3500	1622	4600	重 庆
924	37	13.36	21.99	949	811	433	四 川
1465		17.91	48.90	1538	1456	120	贵 州
469	87	4.85	40.77	245	144	976	云 南
						3	西 藏
32127	59	78.71	85.43	2847	2638	3942	陕 西
496	2	31.24	72.68	758	486	920	甘 肃
							青 海
305		9.11	17.95	723	309	66	宁 夏
81		8.36	13.51	244	1	7665	新 疆

2-8-1 全国历年县城燃气情况
National County Seat Gas in Past Years

年 份 Year	人工煤气 Man-Made Coal Gas				天然气 Natural Gas			
	供气总量 (亿立方米) Total Gas Supplied (100 million m³)	家庭用量 Domestic Consumption	用气人口 (万人) Population with Access to Gas (10,000 persons)	管道长度 (公里) Length of Gas Supply Pipeline (km)	供气总量 (亿立方米) Total Gas Supplied (100 million m³)	家庭用量 Domestic Consumption	用气人口 (万人) Population with Access to Gas (10,000 persons)	管道长度 (公里) Length of Gas Supply Pipeline (km)
2000	1.72	1.63	73.10	615	3.31	2.25	237	5268
2001	2.14	1.85	89.85	424	4.37	2.71	274	6400
2002	1.19	1.13	68.30	493	6.36	3.30	316	7398
2003	0.73	0.60	51.33	519	7.67	4.16	361	8397
2004	1.80	1.51	81.91	551	10.97	5.43	437	9881
2005	3.04	2.01	127.35	830	18.12	5.83	519	12602
2006	1.26	0.49	54.58	745	16.47	7.11	780	17487
2007	1.44	0.51	58.52	1158	24.45	7.01	943	21882
2008	2.68	1.84	72.22	1426	23.26	9.00	1123	27110
2009	1.78	0.97	69.12	1459	32.16	13.79	1404	34214
2010	4.06	1.03	69.97	1520	39.98	17.09	1835	42156
2011	9.51	1.14	66.53	1458	53.87	22.83	2414	52450
2012	8.57	1.08	53.58	1255	70.14	28.11	2926	66697
2013	7.65	2.24	63.15	1345	81.58	31.55	3555	77122
2014	8.49	2.12	56.05	1542	92.65	34.50	4165	88862
2015	8.21	2.29	55.50	1376	102.60	37.97	4715	106466
2016	7.18	1.03	60.08	1366	105.70	39.25	5096	105966
2017	7.41	1.38	65.65	1296	137.96	48.00	6186	126539
2018	6.22	3.31	68.90	1869	171.04	57.61	6848	144314
2019	3.62	1.69	70.20	2671	201.87	67.21	7520	167650
2020	4.17	1.92	69.50	2761	214.53	72.13	8170	186244

2-8-1　续表　continued

| 年 份

Year | 液化石油气　LPG | | | | 燃气普及率
(%)
Gas Coverage Rate

(%) |
	供气总量 (万吨) Total Gas Supplied (10,000 tons)	家庭用量 Domestic Consumption	用气人口 (万人) Population with Access to Gas (10,000 persons)	管道长度 (公里) Length of Gas Supply Pipeline (km)	
2000	110.84	98.23	2723	96	54.41
2001	127.55	109.90	3653	674	44.55
2002	142.42	124.23	4025	760	49.69
2003	174.45	140.17	4508	1033	53.28
2004	188.94	156.37	4961	1172	56.87
2005	185.90	147.10	5151	1203	57.80
2006	195.04	147.07	5405	1728	52.45
2007	203.22	158.60	6217	2355	57.33
2008	202.14	160.60	6504	2899	59.11
2009	212.58	171.00	6776	3136	61.66
2010	218.50	174.97	7098	3053	64.89
2011	242.17	205.23	7058	2594	66.52
2012	256.94	212.13	7241	2773	68.50
2013	241.07	196.89	7208	2236	70.91
2014	235.32	198.38	7242	2538	73.24
2015	230.01	192.99	7081	2068	75.90
2016	219.22	184.36	6919	1579	78.19
2017	215.48	183.50	6458	1504	81.35
2018	214.06	181.31	6243	1829	83.85
2019	217.10	177.81	6128	1865	86.47
2020	199.54	170.15	5874	1461	89.07

2-8-2 县城人工煤气(2020年)

地区名称 Name of Regions	生产能力 (万立方米／日) Production Capacity (10,000 m³)	储气能力 (万立方米) Gas Storage Capacity (10,000 m³)	供气管道 长　度 (公里) Length of Gas Supply Pipeline (km)	自制气量 (万立方米) Self-Produced Gas (10,000 m³)	合　计 Total
全　国　National Total	35.00	614.92	2760.61	3000	41714.08
河　北　Hebei		7.00	690.31		13617.52
山　西　Shanxi	35.00	25.00	2020.80	3000	24011.15
内 蒙 古　Inner Mongolia					
辽　宁　Liaoning					
吉　林　Jilin					
黑 龙 江　Heilongjiang					
江　苏　Jiangsu					
浙　江　Zhejiang					
安　徽　Anhui					
福　建　Fujian					
江　西　Jiangxi					
山　东　Shandong		3.00			3068.00
河　南　Henan					
湖　北　Hubei					
湖　南　Hunan					
广　东　Guangdong					
广　西　Guangxi					
海　南　Hainan					
重　庆　Chongqing					
四　川　Sichuan		390.00			390.00
贵　州　Guizhou					
云　南　Yunnan		189.42			0.41
西　藏　Tibet					
陕　西　Shaanxi					
甘　肃　Gansu					
青　海　Qinghai					
宁　夏　Ningxia					
新　疆　Xinjiang		0.50	49.50		627.00

County Seat Man-Made Coal Gas(2020)

供气总量(万立方米) Total Gas Supplied (10,000 m³)		燃气损失量	用气户数 (户)	家庭用户 Household User	用气人口 (万人)	地区名称
销售气量 Quantity Sold	居民家庭 Households	燃气损失量 Loss Amount	Number of Households with Access to Gas (unit)	家庭用户 Household User	Population with Access to Gas (10,000 persons)	Name of Regions
40588.79	**19248.29**	**1125.29**	**280166**	**275367**	**69.50**	全 国
13566.84	3757.86	50.68	56995	56707	18.99	河 北
22996.55	14786.34	1014.60	202289	199213	46.04	山 西
						内 蒙 古
						辽 宁
						吉 林
						黑 龙 江
						江 苏
						浙 江
						安 徽
						福 建
						江 西
3010.00		58.00	37			山 东
						河 南
						湖 北
						湖 南
						广 东
						广 西
						海 南
						重 庆
390.00	390.00		3640	3640	1.10	四 川
						贵 州
0.40	0.09	0.01	7216	6228	2.56	云 南
						西 藏
						陕 西
						甘 肃
						青 海
						宁 夏
625.00	314.00	2.00	9989	9579	0.81	新 疆

2-8-3 县城天然气(2020年)

地区名称 Name of Regions	储气能力 (万立方米) Gas Storage Capacity (10,000 m³)	供气管道 长 度 (公里) Length of Gas Supply Pipeline (km)	供气总量(万立方米)			
			合 计 Total	销售气量 Quantity Sold	居民家庭 Households	集中供热 Central Heating
全 国 National Total	13685.50	186243.63	2145337.47	2111506.66	721334.18	152320.78
河 北 Hebei	1148.43	21790.68	209961.99	206820.67	91428.43	25760.81
山 西 Shanxi	533.50	10063.37	199978.92	197416.68	48645.14	16203.60
内 蒙 古 Inner Mongolia	312.37	3396.58	57927.27	57014.50	12739.27	4991.29
辽 宁 Liaoning	147.70	2664.41	27244.06	26713.77	4368.28	624.38
吉 林 Jilin	343.18	1230.03	8194.53	8103.36	2992.10	252.00
黑 龙 江 Heilongjiang	184.03	1537.80	10912.77	10752.72	4203.07	1034.67
江 苏 Jiangsu	427.67	8377.94	72864.71	71986.91	31485.41	1720.00
浙 江 Zhejiang	604.12	7515.14	125076.60	123944.24	9711.25	
安 徽 Anhui	572.85	12541.18	110887.45	108676.73	33845.73	58.90
福 建 Fujian	702.87	4884.90	48769.38	47976.38	6266.99	
江 西 Jiangxi	755.43	7450.52	84075.15	83396.34	17948.84	1436.52
山 东 Shandong	625.49	19478.14	231025.60	228981.56	63093.18	12286.26
河 南 Henan	549.94	10957.69	160739.86	158198.83	63776.00	5656.46
湖 北 Hubei	191.73	7128.50	41687.53	40971.39	19413.92	136.00
湖 南 Hunan	1753.92	10335.70	66780.96	65671.83	32961.78	619.25
广 东 Guangdong	344.82	3121.15	24198.15	24001.21	9144.85	
广 西 Guangxi	275.27	1848.07	14184.18	14040.71	3767.73	
海 南 Hainan	121.70	918.46	5123.63	5065.21	1217.73	
重 庆 Chongqing	69.03	2847.71	27785.34	27026.21	18087.60	
四 川 Sichuan	224.34	24637.58	195764.16	187932.92	115212.04	89.10
贵 州 Guizhou	702.52	3691.53	24558.47	24344.95	5696.38	126.13
云 南 Yunnan	661.25	2965.00	18597.09	18315.51	4357.20	137.00
西 藏 Tibet			3.20	3.00	1.50	
陕 西 Shaanxi	1216.06	7044.67	134143.41	132377.86	44166.73	39065.36
甘 肃 Gansu	492.21	2015.04	40612.02	40153.81	9114.07	6820.37
青 海 Qinghai	56.30	693.21	30862.70	30253.67	11715.36	9417.67
宁 夏 Ningxia	31.95	1772.18	45043.28	44752.71	14137.24	3232.68
新 疆 Xinjiang	636.82	5336.45	128335.06	126612.98	41836.36	22652.33

County Seat Natural Gas(2020)

Total Gas Supplied (10,000 m³)		用气户数 (户) Number of Household with Access to Gas (unit)	家庭用户 Household User	用气人口 (万人) Population with Accessto Gas (10,000 persons)	天 然 气 汽车加气站 (座) Gas Stations for CNG-Fueled Motor Vehicles (unit)	地区名称 Name of Regions
燃气汽车 Gas-Powered Automobiles	燃气损失量 Loss Amount					
345487.84	33830.81	27997794	26599281	8170.08	1745	全 国
20587.84	3141.32	2719344	2637732	772.91	160	河 北
81179.14	2562.24	1252943	1236656	375.04	113	山 西
22307.26	912.77	559825	528112	165.50	134	内 蒙 古
6718.30	530.29	345469	336993	96.39	51	辽 宁
3942.96	91.17	292049	283611	74.90	29	吉 林
4145.46	160.05	287371	276222	80.72	63	黑 龙 江
6647.23	877.80	1325046	1291242	369.11	38	江 苏
1465.84	1132.36	760706	734048	221.17	11	浙 江
20877.02	2210.72	1863781	1741771	563.79	70	安 徽
979.09	793.00	561703	550840	183.88	11	福 建
740.99	678.81	896548	868469	301.44	8	江 西
15144.08	2044.04	2764588	2707780	842.84	148	山 东
18120.08	2541.03	2562208	2467615	851.75	142	河 南
2714.22	716.14	830176	808680	233.77	32	湖 北
6892.86	1109.13	1633405	1472777	511.13	39	湖 南
423.24	196.94	398802	376687	117.82	7	广 东
234.56	143.47	226046	221267	75.53	2	广 西
2078.12	58.42	226676	143903	19.60	6	海 南
3024.16	759.13	771090	741459	183.95	15	重 庆
17720.10	7831.24	3788188	3603930	971.83	69	四 川
2647.05	213.52	376882	355521	154.52	26	贵 州
889.36	281.58	210245	203940	84.37	19	云 南
	0.20	1300	912	0.40		西 藏
20159.30	1765.55	1294205	1134786	350.03	152	陕 西
20575.29	458.21	404445	358803	145.24	81	甘 肃
2347.76	609.03	159743	141447	48.24	14	青 海
10316.85	290.57	343321	327434	64.86	53	宁 夏
52609.68	1722.08	1141689	1046644	309.35	252	新 疆

2-8-4 县城液化石油气(2020年)

地区名称 Name of Regions		储气能力 (吨) Gas Storage Capacity (ton)	供气管道长度 (公里) Length of Gas Supply Pipeline (km)	供气总量(吨)		
				合计 Total	销售气量 Quantity Sold	居民家庭 Households
全　国	National Total	318687.19	1461.38	1995410.63	1977974.84	1701549.46
河　北	Hebei	11673.80	199.36	77555.00	76642.07	68200.69
山　西	Shanxi	9193.65	32.72	41107.01	40703.72	21980.42
内 蒙 古	Inner Mongolia	8812.00	59.97	46738.21	46354.82	39561.66
辽　宁	Liaoning	5745.50	30.85	30324.75	30171.77	27128.37
吉　林	Jilin	2922.40	81.08	21779.08	21677.50	18281.05
黑 龙 江	Heilongjiang	5846.10	9.60	30822.30	30488.97	27487.45
江　苏	Jiangsu	18310.50	0.80	90444.47	89518.38	64742.96
浙　江	Zhejiang	13788.56	47.71	152544.67	151894.38	121104.23
安　徽	Anhui	18325.66	31.29	110922.86	110392.09	98178.18
福　建	Fujian	10674.59	17.78	89201.07	88766.51	78834.90
江　西	Jiangxi	22983.70	155.55	165565.94	164145.33	147517.71
山　东	Shandong	17845.10	143.83	77723.16	77148.22	64855.08
河　南	Henan	15489.67	1.55	158419.93	155886.45	135672.40
湖　北	Hubei	9759.55	19.17	53558.43	53015.30	50627.04
湖　南	Hunan	40765.96	333.34	225210.56	223265.10	197817.20
广　东	Guangdong	17383.85	82.01	187333.23	186341.99	156868.82
广　西	Guangxi	11448.96	0.50	140203.24	138619.75	131454.22
海　南	Hainan	1470.50		24209.10	23975.49	21984.74
重　庆	Chongqing	5313.00		13804.34	13705.93	10348.79
四　川	Sichuan	10084.32	12.64	28384.50	27983.50	23790.39
贵　州	Guizhou	8345.85		58156.65	57718.35	47314.84
云　南	Yunnan	19867.00	107.00	60053.01	59512.28	52666.40
西　藏	Tibet	2707.60	21.55	13618.16	13558.95	12306.47
陕　西	Shaanxi	13682.77	4.10	36117.57	35665.61	33415.49
甘　肃	Gansu	5327.29	15.53	25926.87	25554.44	22323.92
青　海	Qinghai	2174.20	0.08	4309.83	4229.17	3366.48
宁　夏	Ningxia	3578.31	0.06	7542.74	7453.44	6085.74
新　疆	Xinjiang	5166.80	53.31	23833.95	23585.33	17633.82

County Seat LPG Supply (2020)

Total Gas Supplied (ton)		用气户数 （户） Number of Household with Access to Gas (unit)	家庭用户 Household User	用气人口 （万人） Population with Access to Gas (10,000 persons)	液化石油气 汽车加气站 （座） Gas Stations for LPG- Fueled Motor Vehicles (unit)	地区 名称 Name of Regions
燃气汽车 Gas-Powered Automobiles	燃气损失量 Loss Amount					
40624.23	17435.79	17668396	16827512	5874.10	164	全　国
0.50	912.93	807273	772078	241.21	5	河　北
15000.00	403.29	327502	308619	113.38	6	山　西
3109.40	383.39	886339	843590	293.40	14	内 蒙 古
413.40	152.98	409102	367854	90.13	12	辽　宁
940.00	101.58	232407	213769	80.26	6	吉　林
1948.54	333.33	494237	450483	144.03	18	黑 龙 江
1080.00	926.09	619829	559189	154.79	1	江　苏
	650.29	932078	834448	227.93		浙　江
320.00	530.77	892559	855077	283.56	8	安　徽
	434.56	781491	763983	291.79		福　建
1201.00	1420.61	1339412	1317971	493.00	3	江　西
	574.94	549049	522863	172.51		山　东
7271.00	2533.48	1154797	1117638	426.17	19	河　南
1225.01	543.13	587010	509317	194.34	4	湖　北
698.15	1945.46	1920036	1885967	589.88	6	湖　南
	991.24	1073008	1058499	372.80	2	广　东
	1583.49	1278190	1246779	452.00		广　西
70.00	233.61	204737	202524	48.85	1	海　南
	98.41	98734	85843	22.84		重　庆
382.00	401.00	229100	210527	86.67	7	四　川
	438.30	939709	904158	403.51	4	贵　州
	540.73	728291	688719	268.54		云　南
502.00	59.21	116233	106766	49.66	5	西　藏
80.00	451.96	318776	306596	114.26	2	陕　西
1879.56	372.43	394826	359789	162.88	12	甘　肃
	80.66	55307	47909	21.93		青　海
1278.46	89.30	68857	67755	22.68	2	宁　夏
3225.21	248.62	229507	218802	51.10	27	新　疆

2-9-1 全国历年县城集中供热情况

年 份 Year	供 热 能 力 Heating Capacity		供热总量 Total Heat Supplied	
	蒸 汽 (吨/小时) Steam (ton/hour)	热 水 (兆瓦) Hot Water (mega watts)	蒸 汽 (万吉焦) Steam (10,000 gigajoules)	热 水 (万吉焦) Hot Water (10,000 gigajoules)
2000	4418	11548	1409	9076
2001	3647	11180	1872	20568
2002	4848	14103	2627	27245
2003	5283	19446	2871	22286
2004	5524	20891	3194	21847
2005	8837	20835	8781	18736
2006	9193	26917	8520	23535
2007	13461	35794	15780	39071
2008	12370	44082	14708	75161
2009	16675	62330	12519	56013
2010	15091	68858	16729	103005
2011	14738	81348	8475	63264
2012	13914	97281	8358	51943
2013	13285	107498	6413	68193
2014	13011	129447	5693	63928
2015	13680	125788	10957	96566
2016	10206	130430	5130	67488
2017	14853	137222	8106	148237
2018	16790	139943	8585	76441
2019	17466	153330	9810	78592
2020	18085	158186	9958	84808

注：2000年蒸汽供热总量计量单位为万吨。

National County Seat Centralized Heating in Past Years

管道长度 Length of Pipelines		集中供热 面　积 (亿平方米) Heated Area	年　份
蒸　汽 (公里) Steam (km)	热　水 (公里) Hot Water (km)	(100 million m²)	Year
1144	4187	0.67	2000
658	4478	0.92	2001
881	4778	1.45	2002
904	6136	1.73	2003
991	7094	1.72	2004
1176	8048	2.06	2005
1367	9450	2.37	2006
1564	12795	3.17	2007
1612	14799	3.74	2008
1874	18899	4.81	2009
1773	23737	6.09	2010
1665	28577	7.81	2011
1965	31901	9.05	2012
2929	37169	10.33	2013
2733	41209	11.42	2014
3283	43013	12.31	2015
2767	44168	13.12	2016
	60793	14.63	2017
	66831	16.18	2018
	75068	17.48	2019
	81366	18.57	2020

Note: Heating capacity through steam in 2000 is measured with the unit of 10,000 tons.

2-9-2　县城集中供热(2020年)

地区名称 Name of Regions	蒸汽　Steam						供热能力 (兆瓦) Heating Capacity (mega watts)	热电厂 供　热 Heating by Co- Generation
	供热能力 (吨/小时) Heating Capacity (ton/hour)	热电厂 供　热 Heating by Co- Generation	锅炉房 供　热 Heating by Boilers	供热总量 (万吉焦) Total Heat Supplied (10,000 gcal)	热电厂 供　热 Heating by Co- Generation	锅炉房 供　热 Heating by Boilers		
全　国 National Total	18085	14800	1979	9958	7887	1106	158186	61828
河　北 Hebei	2522	1944	484	1293	940	163	27981	5238
山　西 Shanxi	2271	2089	182	1346	1060	286	20611	8741
内 蒙 古 Inner Mongolia	2101	1966	135	1033	902	131	27333	14371
辽　宁 Liaoning	1395	945	450	710	535	115	13612	9162
吉　林 Jilin				221			6389	288
黑 龙 江 Heilongjiang	2977	2391	350	1855	1624	179	11278	5072
江　苏 Jiangsu	110			11				
浙　江 Zhejiang								
安　徽 Anhui	60	20	40	35	5	30	90	90
福　建 Fujian								
江　西 Jiangxi								
山　东 Shandong	4849	4579	250	2400	2232	168	13497	11001
河　南 Henan	1214	320	29	505	197	17	3414	2445
湖　北 Hubei								
湖　南 Hunan								
广　东 Guangdong								
广　西 Guangxi								
海　南 Hainan								
重　庆 Chongqing								
四　川 Sichuan				4			90	8
贵　州 Guizhou								
云　南 Yunnan								
西　藏 Tibet	40		40	7		6	314	92
陕　西 Shaanxi	547	547		393	393		3683	1234
甘　肃 Gansu				145			10090	1308
青　海 Qinghai			19			10	1968	195
宁　夏 Ningxia							4655	387
新　疆 Xinjiang							13183	2197

County Seat Central Heating(2020)

| 热水　Hot Water | | | | 管道长度
(公里)
Length of
Pipelines
(km) | 一级管网
First
Class | 二级管网
Second
Class | 供热面积
(万平方米)
Heated Area
(10,000 m²) | 住宅
Housing | 公共建筑
Public
Building | 地区名称
Name of
Regions | |
锅炉房 供　热 Heating By Boilers	供热总量 (万吉焦) Total Heat Supplied (10,000gcal)	热电厂 供　热 Heating by Co- Generation	锅炉房 供　热 Heating by Boilers								
78826	**84808**	**26880**	**49248**	**81366**	**29633**	**51733**	**185655**	**134565**	**34478**	全	国
12661	13452	2964	8399	14350	5244	9106	32496	24888	4552	河	北
9523	12547	3573	5620	11012	3463	7549	26113	19517	4487	山	西
12242	13388	4912	8027	11548	4236	7312	25444	17577	5460	内 蒙 古
4094	4501	1636	2603	4921	1316	3604	9332	6354	1892	辽	宁
6101	4298	191	4107	4751	1477	3274	9069	6454	2248	吉	林
5629	7585	3340	3869	6121	1883	4238	16923	12409	4193	黑 龙 江
	76			93	93		193	193		江	苏
										浙	江
	8	8		141	41	101	839	58	8	安	徽
										福	建
										江	西
1416	7683	6160	1070	12532	5006	7527	23456	19792	2555	山	东
468	1867	519	359	2302	873	1429	4901	3324	789	河	南
										湖	北
										湖	南
										广	东
										广	西
										海	南
										重	庆
59	36	3	20	153	48	105	87	21	66	四	川
										贵	州
										云	南
55	122	40	46	465	168	297	231	70	132	西	藏
1959	2100	923	962	1177	636	541	4803	3281	408	陕	西
8777	6806	948	5855	3326	1545	1781	12475	8489	2394	甘	肃
1747	1076	58	993	1073	531	542	2012	1182	830	青	海
4188	2443	275	2168	2290	1239	1050	4563	3362	610	宁	夏
9908	6820	1330	5152	5111	1834	3277	12716	7593	3855	新	疆

三、居民出行数据
Data by Residents Travel

2-10-1 全国历年县城道路和桥梁情况
National County Seat Road and Bridge in Past Years

年 份 Year	道路长度 (万公里) Length of Roads (10,000 km)	道路面积 (亿平方米) Surface Area of Roads (100 million m²)	防洪堤长度 (万公里) Length of Flood Control Dikes (10,000 km)	人均城市道路面积 (平方米) Urban Road Surface Area Per Capita (m²)
2000	5.04	6.24	0.93	11.20
2001	5.10	7.67	0.90	8.51
2002	5.32	8.31	0.96	9.37
2003	5.77	9.06	0.93	9.82
2004	6.24	9.92	1.00	10.30
2005	6.68	10.83	0.98	10.80
2006	7.36	12.26	1.32	10.30
2007	8.38	13.44	1.17	10.70
2008	8.88	14.60	1.33	11.21
2009	9.50	15.98	1.19	11.95
2010	10.59	17.60	1.25	12.68
2011	10.86	19.24	1.37	13.42
2012	11.80	21.02	1.38	14.09
2013	12.52	22.69		14.86
2014	13.04	24.08		15.39
2015	13.35	24.95		15.98
2016	13.16	25.35		16.41
2017	14.08	26.84		17.18
2018	14.48	27.82		17.73
2019	15.16	29.01		18.29
2020	15.94	29.97		18.92

注: 1.自2006年起, 人均道路面积按县城人口和县城暂住人口合计为分母计算。
  2.自2013年起, 不再统计防洪堤长度数据。

Note: 1.Since 2006, road surface per capita has been calculated based on denominator which combines both permanent and temporary residents in county seat areas.
  2.Starting from 2013, the data on the length of flood prevention dyke has been unavailable.

2-10-2 县城道路和桥梁(2020年)
County Seat Roads and Bridges(2020)

地区名称 Name of Regions	道路长度 (公里) Length of Roads (km)	建成区 in Built District	道路面积 (万平方米) Surface Area of Roads (10,000 m²)	人行道 面 积 Surface Area of Sidewalks	建成区 in Built District	桥梁数 (座) Number of Bridges (unit)
全　国　National Total	**159436.13**	**143609.36**	**299745.22**	**74764.56**	**274386.60**	**17298**
河　北　Hebei	12633.72	12256.59	25637.29	7171.99	24703.06	774
山　西　Shanxi	5491.56	5256.16	10533.69	2577.60	10008.75	526
内 蒙 古　Inner Mongolia	7199.58	6824.24	15165.49	4132.60	14463.29	329
辽　宁　Liaoning	1934.07	1664.97	3289.41	870.12	2796.43	221
吉　林　Jilin	1484.41	1424.80	2584.46	712.80	2514.97	148
黑 龙 江　Heilongjiang	4165.85	4025.00	5022.68	1103.60	4718.19	280
江　苏　Jiangsu	5430.07	4623.34	11249.25	2334.71	9394.78	793
浙　江　Zhejiang	6481.86	5591.41	10885.55	2500.28	9687.72	1575
安　徽　Anhui	9138.03	8115.45	21026.15	5091.08	18388.36	1311
福　建　Fujian	5682.50	4941.67	8953.25	2027.95	8085.64	594
江　西　Jiangxi	9106.03	8302.46	17516.28	4102.56	15605.10	711
山　东　Shandong	11317.35	9633.93	22848.95	4704.30	20667.62	1614
河　南　Henan	11976.89	10837.62	27669.25	6553.44	24961.55	1754
湖　北　Hubei	4057.23	3824.93	8364.70	2248.41	7899.94	458
湖　南　Hunan	10889.20	9121.20	17499.54	4753.30	16546.09	562
广　东　Guangdong	4641.73	4013.86	6808.05	1745.40	5997.39	358
广　西　Guangxi	6044.13	5905.25	11211.55	2721.74	10903.19	748
海　南　Hainan	1004.80	748.99	1863.27	466.59	1559.08	78
重　庆　Chongqing	1226.31	1200.48	2176.05	669.83	2172.42	285
四　川　Sichuan	8553.41	7995.69	16359.62	4527.61	15503.04	1109
贵　州　Guizhou	8005.35	6683.59	13110.82	3189.79	11313.57	709
云　南　Yunnan	5880.26	5403.58	10650.07	2761.04	10021.21	682
西　藏　Tibet	1117.67	802.32	1340.63	398.86	957.41	133
陕　西　Shaanxi	4724.46	4335.94	8063.46	2367.53	7536.16	544
甘　肃　Gansu	3218.14	2888.03	5690.94	1512.31	4955.76	464
青　海　Qinghai	1523.93	1401.33	2506.31	612.04	2425.19	150
宁　夏　Ningxia	1393.79	1313.96	2919.26	935.16	2775.78	76
新　疆　Xinjiang	5113.80	4472.57	8799.25	1971.92	7824.91	312

2-10-2 续表 continued

地区名称 Name of Regions	大桥及 特大桥 Great Bridge and Grand Bridge	立交桥 Intersection	道路照明 灯盏数 （盏） Number of Road Lamps (unit)	安装路灯 道路长度 （公里） Length of The Road with Street Lamp (km)	地下综合 管廊长度 （公里） Length of The Utility Tunnel (km)	新建地下综 合管廊长度 （公里） Length of The New-built Utility Tunnel (km)
全 国 National Total	2163	444	8965336	114568	1041.05	468.84
河 北 Hebei	101	57	577002	8984	78.90	10.16
山 西 Shanxi	103	37	309024	3623	21.91	23.41
内 蒙 古 Inner Mongolia	37	11	391064	4928	30.10	15.30
辽 宁 Liaoning	5	3	129276	1387	2.80	0.30
吉 林 Jilin	21	8	113458	975		
黑 龙 江 Heilongjiang	25	19	166621	2440		
江 苏 Jiangsu	151	14	394281	4745	6.25	6.25
浙 江 Zhejiang	140	32	303620	4990	3.13	2.06
安 徽 Anhui	50	27	544474	7479	99.73	3.31
福 建 Fujian	86	23	350840	3931	79.09	9.54
江 西 Jiangxi	147	25	546905	6964	13.81	15.85
山 东 Shandong	58	38	569055	8015	45.74	13.85
河 南 Henan	48	27	625050	8699	6.95	6.45
湖 北 Hubei	83	9	203902	3749	28.96	13.26
湖 南 Hunan	168	15	552043	7495	5.22	5.22
广 东 Guangdong	72		310840	3978		
广 西 Guangxi	54	8	395875	4160	1.16	1.16
海 南 Hainan	9		56422	638		
重 庆 Chongqing	54	9	123374	1105	24.61	21.11
四 川 Sichuan	247	13	583358	6545	230.25	150.87
贵 州 Guizhou	86	9	434784	4118	151.53	81.01
云 南 Yunnan	81	8	411098	4489	132.05	36.94
西 藏 Tibet	18	4	32919	648	72.15	46.28
陕 西 Shaanxi	190	29	264362	3243	3.37	4.64
甘 肃 Gansu	76	17	155045	2300	0.47	
青 海 Qinghai	14		60763	703	1.87	1.87
宁 夏 Ningxia	22		72346	994		
新 疆 Xinjiang	17	2	287535	3246	1.00	

四、环境卫生数据
Data by Environmental Health

2-11-1　全国历年县城排水和污水处理情况
National County Seat Drainage and Wastewater Treatment in Past Years

年　份 Year	排水管道长度 (万公里) Length of Drainage Pipelines (10,000 km)	污水年排放量 (亿立方米) Annual Quantity of Wastewater Discharged (100 million m³)	污水处理厂 Wastewater Treatment Plant		污 水 年 处理总量 (亿立方米) Annual Treatment Capacity (100 million m³)	污水处理率 (%) Wastewater Treatment Rate (%)
			座数 (座) Number of Wastewater Treatment Plant (unit)	处理能力 (万立方米／日) Treatment Capacity (10,000 m³/day)		
2000	4.00	43.20	54	55	3.26	7.55
2001	4.40	40.14	54	455	3.31	8.24
2002	4.44	43.58	97	310	3.18	11.02
2003	5.32	41.87	93	426	4.14	9.88
2004	6.01	46.33	117	273	5.20	11.23
2005	6.04	47.40	158	357	6.75	14.23
2006	6.86	54.63	204	496	6.00	13.63
2007	7.68	60.10	322	725	14.10	23.38
2008	8.39	62.29	427	961	19.70	31.58
2009	9.63	65.70	664	1412	27.36	41.64
2010	10.89	72.02	1052	2040	43.30	60.12
2011	12.18	79.52	1303	2409	55.99	70.41
2012	13.67	85.28	1416	2623	62.18	75.24
2013	14.88	88.09	1504	2691	69.13	78.47
2014	16.03	90.47	1555	2882	74.29	82.12
2015	16.79	92.65	1599	2999	78.95	85.22
2016	17.19	92.72	1513	3036	81.02	87.38
2017	18.98	95.07	1572	3218	85.77	90.21
2018	19.98	99.43	1598	3367	90.64	91.16
2019	21.34	102.30	1669	3587	95.71	93.55
2020	22.39	103.76	1708	3770	98.62	95.05

2-11-2 县城排水和污水处理(2020年)

地区名称 Name of Regions	污水排放量 (万立方米) Annual Quantity of Wastewater Discharged (10,000 m³)	排水管道长度 (公里) Length of Drainage Piplines (km)	污水管道 Sewers	雨水管道 Rainwater Drainage Pipeline	雨污合流管道 Combined Drainage Pipeline	建成区 in Built District	污水处理厂			
							座数(座) Number of Wastewater Treatment Plant (unit)	二、三级处理 Secondary and Tertiary Treatment	处理能力(万立方米/日) Treatment Capacity (10,000 m³/day)	二、三级处理 Secondary and Tertiary Treatment
全 国 National Total	1037607	223919	104035	77296	42587	200224	1708	1428	3770.0	3187.8
河 北 Hebei	72910	13916	7256	6086	575	13543	109	97	354.7	320.2
山 西 Shanxi	34108	8165	3937	2397	1831	7449	85	74	129.1	111.0
内蒙古 Inner Mongolia	28680	8981	4485	3013	1482	8414	69	58	106.5	90.4
辽 宁 Liaoning	20995	2455	883	774	798	2253	31	15	81.6	38.3
吉 林 Jilin	12823	2240	1012	911	317	2233	20	10	49.5	27.0
黑龙江 Heilongjiang	19486	3872	1528	730	1614	3673	49	49	76.9	76.9
江 苏 Jiangsu	43475	9534	3383	4874	1277	8696	32	30	134.1	130.6
浙 江 Zhejiang	53764	11351	6622	4053	675	9764	44	44	169.1	169.1
安 徽 Anhui	66786	17729	7858	7826	2046	14693	62	59	235.8	225.8
福 建 Fujian	42408	8110	4034	3309	767	7546	45	44	143.1	142.1
江 西 Jiangxi	47813	13357	6023	4345	2988	12186	73	46	145.6	92.6
山 东 Shandong	79203	16725	7398	7975	1353	15200	83	83	364.5	364.5
河 南 Henan	83093	19035	8066	6420	4550	17320	125	88	402.0	292.2
湖 北 Hubei	36585	5456	2323	1443	1690	5041	41	36	118.6	95.1
湖 南 Hunan	92784	14315	6031	4342	3942	12388	84	55	277.3	162.0
广 东 Guangdong	36642	4540	1543	1294	1703	3798	41	35	106.6	91.9
广 西 Guangxi	39194	8386	3688	2399	2299	8030	66	63	108.9	105.6
海 南 Hainan	4861	1114	570	391	153	579	11	4	15.6	8.8
重 庆 Chongqing	13834	2857	1714	1003	140	2786	22	17	49.0	35.0
四 川 Sichuan	63373	12889	6084	4335	2470	11841	145	106	198.4	153.9
贵 州 Guizhou	28055	7256	4347	2082	827	4765	96	96	85.7	85.7
云 南 Yunnan	34328	10576	5702	3391	1484	9423	101	96	112.7	108.1
西 藏 Tibet	3922	1317	356	235	725	1091	20	14	6.0	4.1
陕 西 Shaanxi	26981	5779	2687	1507	1585	4971	71	68	100.9	98.6
甘 肃 Gansu	15509	4963	2771	1401	792	4421	65	61	59.3	55.8
青 海 Qinghai	5591	1816	783	551	482	1730	36	27	21.5	17.6
宁 夏 Ningxia	7618	1764	163	200	1401	1589	15	14	32.0	31.0
新 疆 Xinjiang	22784	5421	2789	10	2622	4802	67	39	85.4	54.1

County Seat Drainage and Wastewater Treatment(2020)

| Wastewater Treatment Plant | | | | 其他污水处理设施
Other Wastewater
Treatment Facilities | | 污水处理
总 量
(万立方米)
Total
Quantity of
Wastewater
Treated | 再生水
Recycled Water | | | 地区名称 |
处理量 (万立 方米) Quantity of Waste- water Treated (10,000 m³)	二、三 级处理 Secondary and Tertiary Treated	干污泥 产生量 (吨) Quantity of Dry Sludge Produced (ton)	干污泥 处置量 (吨) Quantity of Dry Sludge Treated (ton)	处理能力 (万立方 米／日) Treatment Capacity (10,000 m³/day)	处理量 (万立方米) Quantity of Wastewater Treated (10,000 m³)	(10,000 m³)	生产能力 (万立方米／日) Recycled Water Production Capacity (10,000 m³/day)	利用量 (万立方米) Annual Quantity of Wastewater Recycled and Reused (10,000 m³)	管道长度 (公里) Length of Piplines (km)	Name of Regions
979709	**829505**	**1699187**	**1619994**	**97.1**	**6485**	**986194**	**811.0**	**123212**	**4209**	全 国
71746	65734	134953	132182			71746	171.1	27343	503	河 北
32840	28867	99885	94650	0.3	74	32914	58.3	7658	322	山 西
27600	23674	85466	83599			27600	60.3	9430	1172	内蒙古
20131	10484	34195	32456	0.2	19	20150	9.2	1206	21	辽 宁
12193	7194	17727	17727	2.0	1	12194	3.0	67	32	吉 林
18410	18410	21150	20690	0.3	0	18410	5.3	119	28	黑龙江
39509	38806	67638	63248	4.5	225	39734	58.2	8043	141	江 苏
52308	52308	58202	58202	5.5	76	52384	35.4	6515	104	浙 江
64133	60881	75896	73917	3.8	102	64235	37.3	6988	117	安 徽
40314	40023	60202	60200			40314	3.0	3	2	福 建
43101	26659	64484	63485	4.1	670	43771	1.9	376		江 西
77423	77423	182121	182000			77423	183.1	29656	356	山 东
80605	59489	181656	165083			80605	46.3	10383	192	河 南
34219	27072	35422	35398			34219	8.1	484	3	湖 北
88999	52409	118257	114334	18.1	601	89600	12.4	346	115	湖 南
33639	29034	44341	30720	0.2	4	33643	2.1	49		广 东
34707	33627	16404	15799	26.9	3078	37785				广 西
4349	1992	13806	12001	0.1	2	4351	2.1	103	17	海 南
13706	9529	19306	550			13706	4.1	285	12	重 庆
56329	44386	80329	80230	13.7	1591	57919	17.2	2409	43	四 川
25588	25588	17263	16738	2.3	19	25607	5.2	1070	18	贵 州
32563	31136	44024	43142	7.5	22	32584	4.2	105	50	云 南
1233	829	3497	3450	0.1	0	1233				西 藏
25037	24616	98486	97632	2.2	1	25038	15.9	1012	25	陕 西
14643	13552	41911	41911			14643	15.7	2046	183	甘 肃
5072	4271	7318	7170			5072	1.0	572	41	青 海
7470	7197	19677	18636			7470	7.1	614	73	宁 夏
21843	14314	55571	54847	5.3	1	21844	43.7	6328	638	新 疆

2-12-1 全国历年县城市容环境卫生情况

年 份 Year	生活垃圾　Domestic Garbage			
	清运量 （万吨） Quantity of Collected and Transported (10,000 ton)	无害化处理场 （厂)座数 （座） Number of Harmless Treatment Plants/Grounds (unit)	无害化 处理能力 （吨／日） Harmless Treatment Capacity (ton/day)	无害化 处理量 （万吨） Quantity of Harmlessly Treated (10,000 ton)
2000	5560	358	18493	782.52
2001	7851	489	29300	1551.88
2002	6503	460	31582	1056.61
2003	7819	380	29546	1159.35
2004	8182	295	26032	865.13
2005	9535	203	23049	688.78
2006	6266	124	15245	414.30
2007	7110	137	18785	496.56
2008	6794	211	34983	838.69
2009	8085	286	45430	1220.15
2010	6317	448	69310	1732.51
2011	6743	683	103583	2728.72
2012	6838	848	126747	3690.65
2013	6506	992	151615	4298.28
2014	6657	1129	168131	4766.44
2015	6655	1187	181429	5259.94
2016	6666	1273	190672	5680.47
2017	6747	1300	205417	6139.95
2018	6660	1324	220696	6212.38
2019	6871	1378	246729	6609.78
2020	6810	1428	358319	6691.32

注：自2006年起，生活垃圾填埋场的统计采用新的认定标准，生活垃圾无害化处理数据与往年不可比。

National County Seat Environmental Sanitation in Past Years

粪　便 清运量 （万吨） Volume of Soil Collected and Transported (10,000 ton)	公共厕所 （座） Number of Latrine (unit)	市容环卫 专用车辆 设备总数 （辆） Number of Vehicles and Equipment Designated for Municipal Environmental Sanitation (unit)	每 万 人 拥有公厕 （座） Number of Latrine per 10,000 Population (unit)	年 份 Year
1301	31309	13118	2.21	2000
1709	31893	13472	3.54	2001
1659	31282	13817	3.53	2002
1699	33139	15114	3.59	2003
1256	34104	16144	3.54	2004
1312	34753	17697	3.46	2005
710	34563	17367	2.91	2006
2507	36542	19220	2.90	2007
1151	37718	20947	2.90	2008
759	39618	22905	2.96	2009
811	40818	25249	2.94	2010
751	40096	28045	2.80	2011
649	41588	31164	2.09	2012
553	42217	35096	2.77	2013
532	43159	38913	2.76	2014
489	43480	42702	2.78	2015
420	43582	46278	2.82	2016
	45808	54575	2.93	2017
	49059	61436	3.13	2018
	52046	69083	3.28	2019
	55548	74722	3.51	2020

Note: Since 2006, treatment of domestic garbage through sanitary landfill has adopted new certification standard, so the datas of harmless treatmented garbage are not compared with the past years.

2-12-2 县城市容环境卫生(2020年)

地区名称 Name of Regions	道路清扫保洁面积（万平方米） Surface Area of Roads Cleaned and Maintained (10,000 m²)	机械化 Mecha-nization	生活垃圾						
			清运量（万吨） Collected and Trans-ported (10,000 ton)	处理量（万吨） Volume of Treated (10,000 ton)	无害化处理厂（场)数（座） Number of Harmless Treatment Plants/ Grounds (unit)	卫生填埋 Sanitary Landfill	焚烧 Inciner-ation	其他 other	无害化处理能力（吨／日） Harmless Treatment Capacity (ton/day)
全 国 **National Total**	**285054**	**210661**	**6809.76**	**6762.64**	**1428**	**1227**	**156**	**45**	**358319**
河 北 Hebei	22782	20224	448.11	447.92	90	85	3	2	101437
山 西 Shanxi	12131	9455	307.34	307.16	78	75	2	1	11120
内 蒙 古 Inner Mongolia	14390	10092	236.72	235.51	68	66		2	7572
辽 宁 Liaoning	4016	2412	115.21	115.21	18	18			3580
吉 林 Jilin	3124	2215	81.37	81.37	14	14			2740
黑 龙 江 Heilongjiang	5884	4307	179.42	177.11	45	41	1	3	5710
江 苏 Jiangsu	9131	7921	225.85	225.85	27	11	15	1	15643
浙 江 Zhejiang	9786	6990	250.87	250.87	42	21	15	6	17525
安 徽 Anhui	20559	17957	312.84	312.84	45	24	21		18855
福 建 Fujian	7654	4447	240.61	240.61	40	30	8	2	12343
江 西 Jiangxi	18110	13522	411.11	411.11	49	41	7	1	12953
山 东 Shandong	22000	17931	367.87	367.87	60	26	30	4	25379
河 南 Henan	27332	20726	618.17	613.26	79	72	6	1	19621
湖 北 Hubei	7883	5555	210.71	210.71	37	32	2	3	6546
湖 南 Hunan	16166	12025	467.87	465.21	59	52	5	2	14470
广 东 Guangdong	7646	3321	273.30	272.60	37	32	4	1	12688
广 西 Guangxi	8944	4318	220.84	220.84	52	45	7		7005
海 南 Hainan	2043	1490	42.79	42.79	7	7			1073
重 庆 Chongqing	2753	1992	83.35	83.17	13	11	2		2591
四 川 Sichuan	14711	10292	446.68	445.67	98	83	9	6	15330
贵 州 Guizhou	10389	8412	256.93	231.87	76	62	9	5	8937
云 南 Yunnan	10482	7147	295.43	295.08	94	84	6	4	9571
西 藏 Tibet	975	69	29.47	28.79	64	64			1093
陕 西 Shaanxi	7065	5049	223.08	220.13	70	68	2		9278
甘 肃 Gansu	5609	4278	189.77	189.14	65	63	1	1	6044
青 海 Qinghai	2093	1108	56.97	54.30	29	29			1822
宁 夏 Ningxia	3136	2220	49.28	49.07	12	12			1197
新 疆 Xinjiang	8263	5185	167.82	166.58	60	59	1		6198

County Seat Environmental Sanitation(2020)

Domestic Garbage			无害化处理量（万吨）				公共厕所（座）Number of Latrines	三类以上 Grade III and Above	市容环卫专用车辆设备总数（辆）Number of Vehicles and Equipment Designated for Municipal Environmental Sanitation	地区名称
卫生填埋 Sanitary Landfill	焚烧 Incineration	其他 other	Volume of Harmlessly Treated (10,000 ton)	卫生填埋 Sanitary Landfill	焚烧 Incineration	其他 other	(unit)		(unit)	Name of Regions
256682	**94051**	**7587**	**6691.32**	**4852.96**	**1714.90**	**123.45**	**55548**	**37716**	**74722**	全　　国
96937	3700	800	437.56	380.94	52.35	4.26	4681	3472	6375	河　　北
9960	860	300	307.16	283.59	21.09	2.48	2195	1442	4307	山　　西
7112		460	230.80	215.56	3.70	11.54	4694	1978	3616	内 蒙 古
3580			104.19	100.08	4.12		1174	359	1083	辽　　宁
2740			81.37	55.77	25.59		617	320	1675	吉　　林
5044	200	466	177.11	143.81	19.87	13.43	1823	450	2430	黑 龙 江
3788	11745	110	225.85	35.53	189.77	0.54	2294	1945	2693	江　　苏
6342	10375	808	250.87	121.70	117.49	11.68	1663	1526	2368	浙　　江
5105	13750		312.84	103.47	206.84	2.53	2366	1760	3198	安　　徽
5208	6975	160	240.61	133.12	107.24	0.25	1766	1366	1839	福　　建
8043	4900	10	411.11	300.11	110.97	0.03	2698	2076	5144	江　　西
6069	18580	730	367.87	101.94	250.19	15.73	2201	1923	3474	山　　东
15621	3700	300	585.86	502.78	78.06	5.01	5300	4123	5189	河　　南
5811	410	325	210.71	177.28	29.89	3.54	1118	821	1682	湖　　北
11352	2450	668	465.21	339.55	110.92	14.74	2453	1590	3342	湖　　南
9338	3050	300	272.60	210.03	57.53	5.04	581	482	2320	广　　东
5025	1980		220.84	176.21	44.63		865	711	2425	广　　西
1073			37.37	37.37			310	188	854	海　　南
2091	500		82.52	70.61	11.91		683	610	624	重　　庆
9898	4722	710	445.67	286.95	149.11	9.61	2569	2039	4642	四　　川
5657	2910	370	231.87	162.06	66.26	3.56	1778	1236	3424	贵　　州
6981	1640	950	293.18	249.27	25.65	18.27	4036	3154	3432	云　　南
1093			27.26	27.26			1216	325	584	西　　藏
8378	900		216.57	210.62	5.83	0.12	2749	1896	2763	陕　　西
5820	104	120	189.14	179.89	8.16	1.08	1382	827	1811	甘　　肃
1822			52.17	52.17			628	102	640	青　　海
1197			49.07	37.39	11.67		417	266	525	宁　　夏
5598	600		163.95	157.90	6.05		1291	729	2263	新　　疆

五、绿色生态数据
Data by Green Ecology

2-13-1　全国历年县城园林绿化情况

计量单位：公顷

年　份 Year	建成区绿化 覆盖面积 Built District Green Coverage Area	建 成 区 绿地面积 Built District Area of Green Space	公园绿地 面　　积 Area of Public Recreational Green Space	公园面积 Park Area
2000	142667	85452	31807	15736
2001	138338	94803	35082	69829
2002	148214	102684	38378	73612
2003	169737	119884	44628	28930
2004	193274	137170	50997	33678
2005	210393	151859	56869	32830
2006	247318	185389	59244	39422
2007	288085	219780	70849	54488
2008	317981	249748	79773	51510
2009	365354	285850	92236	56015
2010	412730	330318	106872	67325
2011	465885	385636	121300	80850
2012	519812	436926	134057	85272
2013	566706	482823	144644	96630
2014	599348	520483	155083	105680
2015	617022	542249	163526	114093
2016	633304	559476	170700	122103
2017	686922	610268	185320	138776
2018	711680	631559	191699	146656
2019	757497	672684	207897	158584
2020	784249	700180	213017	167699

注：1.自2006年起，"公共绿地"统计为"公园绿地"。
　　2.自2006年起，"人均公共绿地面积"统计为以城区人口和城区暂住人口合计为分母计算的"人均公园绿地面积"。

National County Seat Landscaping in Past Years

Measurement Unit: Hectare

人均公园 绿地面积 (平方米) Public Recreational Green Space Per Capita (m^2)	建成区绿化 覆盖率 (%) Green Coverage Rate of Built District (%)	建成区 绿地率 (%) Green Space Rate of Built District (%)	年 份 Year
5.71	10.86	6.51	2000
3.88	13.24	9.08	2001
4.32	14.12	9.78	2002
4.83	15.27	10.79	2003
5.29	16.42	11.65	2004
5.67	16.99	12.26	2005
4.98	18.70	14.01	2006
5.63	20.20	15.41	2007
6.12	21.52	16.90	2008
6.89	23.48	18.37	2009
7.70	24.89	19.92	2010
8.46	26.81	22.19	2011
8.99	27.74	23.32	2012
9.47	29.06	24.76	2013
9.91	29.80	25.88	2014
10.47	30.78	27.05	2015
11.05	32.53	28.74	2016
11.86	34.60	30.74	2017
12.21	35.17	31.21	2018
13.10	36.64	32.54	2019
13.44	37.58	33.55	2020

Note: 1.Since 2006, Public Green Space is changed to Public Recreational Green Space.

2.Since 2006, Public recreational green space per capita has been calculated based on denominator which combines both permanent and temporary residents in urban areas.

2-13-2 县城园林绿化(2020年)

地区名称 Name of Regions	绿化覆盖面积 (公顷) Green Coverage Area (hectare)	建成区 Built District	绿地面积 (公顷) Area of Green Space (hectare)	建成区 Built District
全 国 National Total	**975508**	**784249**	**834122**	**700180**
河 北 Hebei	70387	58477	58687	52852
山 西 Shanxi	32250	28911	27452	25580
内 蒙 古 Inner Mongolia	48000	36596	44265	34057
辽 宁 Liaoning	7743	7304	5905	5540
吉 林 Jilin	12151	8045	10253	7096
黑 龙 江 Heilongjiang	18701	16419	15802	14509
江 苏 Jiangsu	49922	29886	41967	27877
浙 江 Zhejiang	34078	26490	30906	23912
安 徽 Anhui	73388	48531	60630	43439
福 建 Fujian	29639	24430	26005	22323
江 西 Jiangxi	52390	45285	45168	40814
山 东 Shandong	79226	62513	66335	55866
河 南 Henan	73840	67006	63413	58451
湖 北 Hubei	28543	21525	23743	18714
湖 南 Hunan	57596	51886	50302	46244
广 东 Guangdong	27951	22498	21690	20098
广 西 Guangxi	32925	27110	27992	24032
海 南 Hainan	6734	6501	4676	4586
重 庆 Chongqing	12253	7567	10759	6862
四 川 Sichuan	55526	50219	48752	45171
贵 州 Guizhou	48054	32709	43529	30147
云 南 Yunnan	31270	28197	27196	24882
西 藏 Tibet	1173	1076	700	654
陕 西 Shaanxi	30442	22665	24982	19933
甘 肃 Gansu	17258	13149	14164	11446
青 海 Qinghai	5368	5184	4736	4392
宁 夏 Ningxia	8746	7411	7756	6799
新 疆 Xinjiang	29955	26657	26356	23904

County Seat Landscaping(2020)

公园绿地 面　　积 （公顷） Area of Public Recreational Green Space (hectare)	公园个数 （个） Number of Parks (unit)	门票免费 Free Parks	公园面积 （公顷） Park Area (hectare)	地区名称 Name of Regions
213017	9624	9128	167699	全　　国
14390	1041	995	12520	河　　北
7306	385	353	6545	山　　西
10707	465	457	10167	内 蒙 古
2446	79	68	1834	辽　　宁
2239	59	58	1837	吉　　林
4875	213	189	3557	黑 龙 江
7622	215	215	4839	江　　苏
6963	461	457	4686	浙　　江
12922	557	541	9900	安　　徽
7406	457	435	5578	福　　建
13114	680	614	11490	江　　西
16896	449	442	12576	山　　东
17492	473	445	11524	河　　南
5570	267	251	4516	湖　　北
13588	469	463	11320	湖　　南
7374	348	340	6716	广　　东
7000	311	279	5860	广　　西
507	49	42	378	海　　南
2893	136	134	2194	重　　庆
16443	509	493	11577	四　　川
9320	351	335	9892	贵　　州
6972	714	660	5763	云　　南
111	53	52	116	西　　藏
5736	261	225	3266	陕　　西
4473	219	198	2794	甘　　肃
877	43	35	266	青　　海
2003	92	92	1922	宁　　夏
5771	268	260	4066	新　　疆

村镇部分

Statistics for Villages and Small Towns

3-1-1 全国历年建制镇及住宅基本情况

年 份 Year	建 制 镇 统计个数 （万个） Number of Towns (10,000 units)	建成区面积 （万公顷） Surface Area of Build Districts (10,000 hectare)	建 成 区 户籍人口 （亿人） Registered Permanent Population (100 million persons)	非农人口 Nonagriculture Population	建 成 区 暂住人口 （亿人） Temporary Population (100 million persons)	本 年 建设投入 （亿元） Construction Input This Year (100 million RMB)
1990	1.01	82.5	0.61	0.28		156
1991	1.03	87.0	0.66	0.30		192
1992	1.20	97.5	0.72	0.32		284
1993	1.29	111.9	0.79	0.34		458
1994	1.43	118.8	0.87	0.38		616
1995	1.50	138.6	0.93	0.42		721
1996	1.58	143.7	0.99	0.42		915
1997	1.65	155.3	1.04	0.44		821
1998	1.70	163.0	1.09	0.46		872
1999	1.73	167.5	1.16	0.49		980
2000	1.79	182.0	1.23	0.53		1123
2001	1.81	197.2	1.30	0.56		1278
2002	1.84	203.2	1.37	0.60		1520
2003						
2004	1.78	223.6	1.43	0.64		2373
2005	1.77	236.9	1.48	0.66		2644
2006	1.77	312.0	1.40		0.24	3013
2007	1.67	284.3	1.31		0.24	2950
2008	1.70	301.6	1.38		0.25	3285
2009	1.69	313.1	1.38		0.26	3619
2010	1.68	317.9	1.39		0.27	4356
2011	1.71	338.6	1.44		0.26	5018
2012	1.72	371.4	1.48		0.28	5751
2013	1.74	369.0	1.52		0.30	7148
2014	1.77	379.5	1.56		0.31	7172
2015	1.78	390.8	1.60		0.31	6781
2016	1.81	397.0	1.62		0.32	6825
2017	1.81	392.6	1.55			7410
2018	1.83	405.3	1.61			7562
2019	1.87	422.9	1.65			8357
2020	1.88	433.9	1.66			9678

National Summary of Towns and Residential Building in Past Years

住 宅 Residential Building	市政公用设施 Public Facilities	本年住宅竣工建筑面积 (亿平方米) Floor Space Completed of Residential Building This Year (100 million m²)	年末实有住宅建筑面积 (亿平方米) Total Floor Space of Residential Buildings (year-end) (100 million m²)	居住人口 (亿人) Resident Population (100 million persons)	人均住宅建筑面积 (平方米) Per Capita Floor Space (m²)	年 份 Year
76	15	0.49	12.3	0.61	19.9	1990
84	19	0.54	12.9	0.65	19.8	1991
115	28	0.62	14.8	0.72	20.5	1992
189	56	0.80	15.8	0.78	20.2	1993
265	79	0.90	17.6	0.85	20.6	1994
305	104	1.00	18.9	0.91	20.7	1995
373	116	1.10	20.5	0.97	21.1	1996
382	122	1.06	21.8	1.01	21.5	1997
402	141	1.09	23.3	1.07	21.8	1998
464	160	1.20	24.8	1.13	22.0	1999
530	185	1.41	27.0	1.19	22.6	2000
575	220	1.47	28.6	1.26	22.7	2001
655	265	1.69	30.7	1.32	23.2	2002
						2003
903	437	1.82	33.7	1.40	24.1	2004
1000	476	1.90	36.8	1.43	25.7	2005
1139	580	2.04	39.1	1.40	27.9	2006
1061	614	1.28	38.9		29.7	2007
1211	726	1.33	41.5		30.1	2008
1465	798	1.47	44.2		32.1	2009
1828	1028	1.67	45.1		32.5	2010
2106	1168	1.72	47.3		33.0	2011
2469	1348	1.97	49.6		33.6	2012
3561	1603	2.65	52.0		34.1	2013
3550	1663	2.66	54.0		34.6	2014
3373	1646	2.53	55.4		34.6	2015
3327	1697	2.32	56.7		34.9	2016
3565	1867	3.00	53.9		34.8	2017
3856	1788	3.32	57.9		36.1	2018
4525	1785	2.75	60.4		36.5	2019
5024	2048	2.79	61.4		37.0	2020

3-1-2 全国历年建制镇市政公用设施情况

年 份 Year	年供水总量 (亿立方米) Annual Supply of Water (100 million m³)	生活用水 Domestic Water Consumption	用水人口 (亿人) Population with Access to Water (100 million persons)	供水普及率 (%) Water Coverage Rate (%)	人均日生 活用水量 (升) Per Capita Daily Water Consumption (liter)	道路长度 (万公里) Length of Roads (10,000 km)
1990	24.4	10.0	0.37	60.1	74.3	7.7
1991	29.5	11.6	0.42	63.9	76.1	8.4
1992	35.0	13.6	0.48	65.8	78.1	9.6
1993	39.5	15.8	0.54	68.5	80.7	10.9
1994	47.1	17.7	0.62	71.5	78.3	12.6
1995	53.7	21.5	0.69	74.2	85.5	13.4
1996	62.2	24.7	0.74	75.0	91.7	15.5
1997	68.4	27.0	0.80	76.6	92.6	17.8
1998	72.8	30.0	0.86	79.1	95.1	18.7
1999	81.4	34.3	0.93	80.2	100.8	19.4
2000	87.7	37.1	0.99	80.7	102.7	21.0
2001	91.4	39.6	1.04	80.3	104.0	22.8
2002	97.3	42.3	1.10	80.4	105.4	24.3
2003						
2004	110.7	49.0	1.20	83.6	112.1	27.5
2005	136.5	54.2	1.25	84.7	118.4	30.1
2006	131.0	44.7	1.17	83.8	104.2	26.0
2007	112.0	42.1	1.19	76.6	97.1	21.6
2008	129.0	45.0	1.27	77.8	97.1	23.4
2009	114.6	46.1	1.28	78.3	98.9	24.5
2010	113.5	47.8	1.32	79.6	99.3	25.8
2011	118.6	49.9	1.37	79.8	100.7	27.4
2012	122.2	51.2	1.42	80.8	99.1	29.1
2013	126.2	53.7	1.49	81.7	98.6	31.0
2014	131.6	55.8	1.55	82.8	98.7	32.7
2015	134.8	57.8	1.60	83.8	98.7	34.5
2016	135.3	59.0	1.64	83.9	99.0	35.9
2017	131.9	59.0	1.48	88.1	109.5	33.5
2018	133.7	58.9	1.55	88.1	104.1	37.7
2019	142.6	61.7	1.63	89.0	103.9	40.9
2020	145.2	64.1	1.64	89.1	107.0	43.9

注： 1.自2006年起，"公共绿地"统计为"公园绿地"。

2.自2006年起，"人均公共绿地面积"统计为以建制镇建成区人口和暂住人口合计为分母计算的"人均公园绿地面积"。

National Municipal Public Facilities of Towns in Past Years

桥梁数 (万座) Number of Bridges (10,000 units)	排水管道 长 度 (万公里) Length of Drainage Piplines (10,000 km)	公园绿地 面 积 (万公顷) Public Green Space (10,000 hectare)	人均公园 绿地面积 (平方米) Public Recreational Green Space Per Capita (m²)	环卫专用 车辆设备 (万辆) Number of Special Vehicles for Environmental Sanitation (10,000 units)	公共厕所 (万座) Number of Latrines (10,000 units)	年 份 Year
2.4	2.7	0.85	1.4	0.5	4.9	1990
2.7	3.2	1.06	1.6	0.7	5.4	1991
3.1	4.0	1.22	1.7	0.8	6.1	1992
3.5	4.8	1.37	1.7	1.1	6.8	1993
3.9	5.5	1.74	2.2	1.2	7.6	1994
4.2	6.2	2.00	2.2	1.6	8.3	1995
4.8	7.5	2.27	2.3	1.7	8.7	1996
5.1	8.1	2.61	2.5	2.0	9.2	1997
5.4	8.8	2.99	2.7	2.3	9.7	1998
5.6	10.0	3.32	2.9	2.5	10.1	1999
6.1	11.1	3.71	3.0	2.9	10.3	2000
6.4	11.9	4.39	3.4	3.2	10.7	2001
6.8	13.0	4.84	3.5	3.3	11.2	2002
						2003
7.2	15.7	6.01	4.2	3.9	11.8	2004
7.7	17.1	6.81	4.6	4.2	12.4	2005
7.2	11.9	3.3	2.4	4.8	9.4	2006
8.3	8.8	2.72	1.8	5.0	9.0	2007
9.1	9.9	3.09	1.9	6.0	12.1	2008
9.9	10.7	3.14	1.9	6.6	11.6	2009
10.0	11.5	3.36	2.0	6.9	9.8	2010
9.7	12.2	3.45	2.0	7.6	10.1	2011
10.4	13.2	3.73	2.1	8.7	10.5	2012
10.8	14.0	4.33	2.4	9.7	14.0	2013
11.0	15.1	4.48	2.4	10.6	11.4	2014
11.4	16.0	4.69	2.5	11.5	11.9	2015
11.1	16.6	4.79	2.5	12.0	11.7	2016
8.4	16.4	5.24	3.1	11.5	12.1	2017
8.3	17.7	4.99	2.8	11.4	11.8	2018
8.4	18.8	4.96	2.7	12.1	12.9	2019
8.1	19.8	5.01	2.7	12.0	13.5	2020

Note: 1.Since 2006, Public Green Space is changed to Public Recreational Green Space.
    2.Since 2006, Public Recreational Green Space Per Capita has been calculated based on denominator which combines both permanent and temporary residents in built area of town.

3-1-3 全国历年乡及住宅基本情况

年 份 Year	乡统计 个 数 （万个） Number of Townships （10,000 units）	建成区 面 积 （万公顷） Surface Area of Build Districts （10,000 hectare）	建 成 区 户籍人口 （亿人） Registered Permanent Population （100 million persons）	非农人口 Nonagriculture Population	建 成 区 暂住人口 （亿人） Temporary Population （100 million persons）	本年建 设投入 （亿元） Construction Input This Year （100 million RMB）
1990	4.02	110.1	0.72	0.17		121
1991	3.90	109.3	0.70	0.16		136
1992	3.72	98.1	0.66	0.15		168
1993	3.64	99.9	0.65	0.15		191
1994	3.39	101.2	0.62	0.14		234
1995	3.42	103.7	0.63	0.15		260
1996	3.15	95.2	0.60	0.14		296
1997	3.03	95.7	0.60	0.14		296
1998	2.91	93.7	0.59	0.15		316
1999	2.87	92.6	0.59	0.15		325
2000	2.76	90.7	0.58	0.14		300
2001	2.35	79.7	0.53	0.14		283
2002	2.26	79.1	0.52	0.14		325
2003						
2004	2.18	78.1	0.53	0.15		344
2005	2.07	77.8	0.52	0.14		377
2006	1.46	92.83	0.35		0.03	355
2007	1.42	75.89	0.34		0.03	352
2008	1.41	81.15	0.34		0.03	438
2009	1.39	75.76	0.33		0.03	471
2010	1.37	75.12	0.32		0.03	558
2011	1.29	74.19	0.31		0.02	535
2012	1.27	79.55	0.31		0.02	634
2013	1.23	73.69	0.31		0.02	706
2014	1.19	72.23	0.30		0.02	671
2015	1.15	70.00	0.29		0.02	559
2016	1.09	67.30	0.28		0.02	524
2017	1.03	63.38	0.25			653
2018	1.02	65.39	0.25			621
2019	0.95	62.95	0.24			665
2020	0.89	61.70	0.24			780

注：2006年以后，本表统计范围由原来的集镇变为乡。

National Summary of Townships and Residential Building in Past Years

住 宅 Residential Building	市政公 用设施 Public Facilities	本年住宅 竣工建筑 面 积 （亿平方米） Floor Space Completed of Residential Building This Year (100 million m^2)	年末实有住 宅建筑面积 （亿平方米） Total Floor Space of Residential Buildings (year-end) (100 million m^2)	居住人口 （亿人） Resident Population (100 million persons)	人均住宅 建筑面积 （平方米） Per Capita Floor Space (m^2)	年 份 Year
61	7	0.52	13.8	0.72	19.1	1990
67	8	0.53	13.8	0.70	19.8	1991
76	10	0.55	13.4	0.65	20.6	1992
85	13	0.49	13.3	0.64	20.6	1993
113	16	0.51	12.8	0.63	20.3	1994
133	22	0.57	12.7	0.62	20.5	1995
151	26	0.59	12.2	0.58	21.0	1996
155	33	0.56	12.3	0.59	21.0	1997
175	37	0.57	12.3	0.58	21.4	1998
193	36	0.66	12.8	0.58	22.1	1999
175	35	0.60	12.6	0.56	22.6	2000
167	33	0.55	12.0	0.52	23.0	2001
188	39	0.57	12.0	0.51	23.6	2002
						2003
188	48	0.56	12.5	0.50	24.9	2004
186	55	0.56	12.8	0.50	25.5	2005
145	66	0.40	9.1	0.35	25.9	2006
147	75	0.26	9.1		27.1	2007
187	99	0.28	9.2		27.2	2008
212	101	0.29	9.4		28.8	2009
262	129	0.35	9.7		29.9	2010
267	122	0.32	9.5		30.3	2011
306	152	0.36	9.6		30.5	2012
365	153	0.39	9.6		31.2	2013
332	132	0.36	9.3		31.2	2014
285	134	0.32	9.0		31.2	2015
260	136	0.29	8.7		31.2	2016
319	175	0.56	7.9		31.5	2017
304	175	0.39	8.4		33.2	2018
300	178	0.44	8.3		33.9	2019
364	171	0.44	8.4		35.4	2020

Note: Since 2006, coverage of the statistics in the table is changed to township.

3-1-4 全国历年乡市政公用设施情况

年 份 Year	年供水总量 (亿立方米) Annual Supply of Water (100 million m³)	生活用水 Domestic Water Consumption	用水人口 (亿人) Population with Access to Water (100 million persons)	供水普及率 (%) Water Coverage Rate (%)	人均日生 活用水量 (升) Per Capita Daily Water Consumption (liter)	道路长度 (万公里) Length of Roads (10,000 km)
1990	10.8	5.0	0.26	35.7	53.4	15.2
1991	12.3	5.1	0.27	39.3	51.1	14.9
1992	12.7	5.2	0.28	42.6	50.6	14.2
1993	12.3	5.6	0.27	40.6	58.2	14.2
1994	12.8	6.0	0.30	47.8	55.5	14.0
1995	13.7	6.4	0.32	49.9	55.4	14.4
1996	13.9	6.6	0.29	49.0	61.1	14.4
1997	16.2	7.3	0.31	52.3	63.7	14.5
1998	17.2	7.9	0.33	55.5	66.4	14.3
1999	17.3	8.8	0.35	58.2	69.7	14.4
2000	16.8	8.8	0.35	60.1	69.2	13.7
2001	15.7	8.2	0.32	61.0	69.3	12.1
2002	16.4	8.4	0.32	62.1	71.9	12.1
2003						
2004	17.4	9.5	0.35	65.8	74.8	12.6
2005	17.5	9.6	0.35	67.2	75.6	12.4
2006	25.8	6.3	0.22	63.4	78.0	7.0
2007	11.9	6.0	0.21	59.1	76.1	6.2
2008	11.9	6.3	0.23	62.6	75.5	6.4
2009	11.4	6.5	0.22	63.5	79.5	6.3
2010	11.8	6.8	0.23	65.6	81.4	6.6
2011	11.5	6.7	0.22	65.7	82.4	6.5
2012	12.0	6.9	0.22	66.7	83.9	6.7
2013	11.5	6.8	0.22	68.2	82.8	6.8
2014	11.3	6.7	0.22	69.3	83.1	7.0
2015	11.2	6.7	0.22	70.4	84.3	7.1
2016	11.2	6.7	0.22	71.9	85.3	7.3
2017	12.6	7.2	0.19	78.8	104.3	6.6
2018	12.1	6.6	0.20	79.2	91.9	8.1
2019	12.9	6.6	0.20	80.5	93.3	8.7
2020	13.5	6.8	0.19	83.9	97.0	8.9

注：1.自2006年起，"公共绿地"统计为"公园绿地"。

2.自2006年起，"人均公共绿地面积"统计为以乡建成区人口和暂住人口合计为分母计算的"人均公园绿地面积"。

National Municipal Public Facilities of Townships in Past Years

桥梁数 (万座) Number of Bridges (10,000 units)	排水管道 长 度 (万公里) Length of Drainage Piplines (10,000 km)	公园绿 地面积 (万公顷) Public Green Space (10,000 hectare)	人均公园 绿地面积 (平方米) Public Recreational Green Space Per Capita (m²)	环卫专用 车辆设备 (万辆) Number of Special Vehicles for Environmental Sanitation (10,000 units)	公共厕所 (万座) Number of Latrines (10,000 units)	年 份 Year
3.3	2.3	0.64	0.88	0.16	5.33	1990
3.5	2.3	0.83	1.19	0.26	5.74	1991
3.7	2.5	0.87	1.32	0.31	5.89	1992
3.4	2.4	0.91	1.4	0.29	5.77	1993
3.3	2.6	1.11	1.76	0.42	6.18	1994
3.4	3.7	1.09	1.73	0.50	6.42	1995
3.4	3.2	1.08	1.79	0.50	6.12	1996
3.5	3.1	1.15	1.91	0.54	6.25	1997
3.5	3.2	1.31	2.22	0.62	6.21	1998
3.5	3.2	1.32	2.23	0.65	6.04	1999
3.4	3.3	1.35	2.33	0.68	5.86	2000
2.9	3.1	1.36	2.56	0.70	5.03	2001
2.9	3.6	1.31	2.54	0.75	5.05	2002
						2003
2.8	4.3	1.41	2.57	0.77	4.58	2004
2.9	4.3	1.37	2.65	0.80	4.57	2005
2.2	1.9	0.29	0.85	0.88	2.92	2006
2.7	1.1	0.24	0.66	1.04	2.76	2007
2.6	1.2	0.26	0.72	1.30	3.34	2008
2.8	1.4	0.30	0.84	1.34	2.96	2009
2.7	1.4	0.31	0.88	1.45	2.75	2010
2.6	1.4	0.30	0.90	1.53	2.58	2011
2.6	1.5	0.32	0.95	1.85	3.08	2012
2.6	1.6	0.35	1.08	2.56	3.94	2013
2.7	1.6	0.34	1.07	2.37	3.19	2014
2.7	1.7	0.34	1.10	2.41	3.04	2015
2.6	1.8	0.33	1.11	2.50	2.99	2016
1.9	1.9	0.40	1.65	2.76	3.18	2017
1.9	2.4	0.37	1.50	2.80	3.55	2018
1.8	2.5	0.38	1.59	2.93	3.91	2019
1.7	2.4	0.40	1.76	2.86	3.88	2020

Note: 1.Since 2006, Public Green Space is changed to Public Recreational Green Space.

2.Since 2006, Public recreational green space per capita has been calculated based on denominator which combines both permanent and temporary residents in build area of township.

3-1-5 全国历年村庄基本情况

年 份 Year	村庄统计 个　数 （万个） Number of Villages (10,000 unit)	村庄现状 用地面积 （万公顷） Area of Villages (10,000 hectare)	村庄户籍 人　口 （亿人） Registered Permanent Population (100 million persons)	非农人口 Non- agriculture Population	村庄暂住 人　口 （亿人） Temporary Population (100 million persons)	本年建设 投　入 （亿元） Construction Input This Year (100 million RMB)	住　宅 Residential Building .
1990	377.3	1140.1	7.92	0.16		662	545
1991	376.2	1127.2	8.00	0.16		744	618
1992	375.5	1187.7	8.06	0.16		793	624
1993	372.1	1202.7	8.13	0.17		906	659
1994	371.3	1243.8	8.15	0.18		1175	885
1995	369.5	1277.1	8.29	0.20		1433	1089
1996	367.6	1336.1	8.18	0.19		1516	1176
1997	365.9	1366.4	8.18	0.20		1538	1175
1998	355.8	1372.6	8.15	0.21		1585	1220
1999	359.0	1346.3	8.13	0.22		1607	1245
2000	353.7	1355.3	8.12	0.24		1572	1203
2001	345.9	1396.1	8.06	0.25		1558	1145
2002	339.6	1388.8	8.08	0.26		2002	1288
2003							
2004	320.7	1362.7	7.95	0.32		2064	1243
2005	313.7	1404.2	7.87	0.31		2304	1374
2006	270.9		7.14		0.23	2723	1524
2007	264.7	1389.9	7.63		0.28	3544	1923
2008	266.6	1311.7	7.72		0.31	4294	2558
2009	271.4	1362.8	7.70		0.28	5400	3456
2010	273.0	1399.2	7.69		0.29	5692	3412
2011	266.9	1373.8	7.64		0.28	6204	3773
2012	267.0	1409.0	7.63		0.28	7420	4312
2013	265.0	1394.3	7.62		0.28	8183	4898
2014	270.2	1394.1	7.63		0.28	8088	5020
2015	264.5	1401.3	7.65		0.28	8203	5059
2016	261.7	1392.2	7.63		0.27	8321	5045
2017	244.9		7.56			9168	5271
2018	245.2		7.71			9830	5355
2019	251.3		7.76			10167	5529
2020	236.3		7.77			11503	5670

National Summary of Villages in Past Years

市政公用 设 施 Public Facilities	本年住宅 竣工建筑 面 积 (亿平方米) Floor Space Completed of Residential Building This Year (100 million m²)	年末实有 住宅建筑 面 积 (亿平方米) Total Floor Space of Residential Buildings (year-end) (100 million m²)	居住人口 (亿人) Resident Population (100 million persons)	人均住宅 建筑面积 (平方米) Per Capita Floor Space (m²)	道路长度 (万公里) Length of Roads (10,000 km)	桥梁数 (万座) Number of Bridges (10,000 units)	年 份 Year
33	4.82	159.3	7.84	20.3	262.1		1990
26	5.54	163.3	7.95	20.5	240.0	37.7	1991
32	4.86	167.4	8.01	20.9	262.9	40.2	1992
57	4.38	170.0	8.12	20.9	268.7	43.0	1993
65	4.49	169.1	7.90	21.4	263.2	43.0	1994
104	4.95	177.7	8.06	22.0	275.0	44.7	1995
106	4.96	182.4	8.13	22.4	279.3	44.1	1996
136	4.66	185.9	8.12	22.9	283.2	44.7	1997
139	4.73	189.2	8.07	23.5	290.3	43.4	1998
152	4.62	192.8	8.04	24.0	287.3	45.7	1999
139	4.47	195.2	8.02	24.3	287.0	46.3	2000
160	4.28	199.1	7.97	25.0	283.6	46.7	2001
368	4.39	202.5	7.94	25.5	287.3	47.1	2002
							2003
342	4.22	205.0	7.75	26.5	285.1	57.8	2004
380	4.42	208.0	7.72	26.9	304.0	58.0	2005
501	4.75	202.9	7.14	28.4	221.9	50.7	2006
616	3.65	222.7		29.2			2007
793	4.10	227.2		29.4			2008
863	4.91	237.0		30.8			2009
1105	4.56	242.6		31.6			2010
1216	4.86	245.1		32.1			2011
1660	5.25	247.8		32.5			2012
1850	5.46	250.6		32.9	228.0		2013
1707	5.46	253.4		33.2	234.1		2014
1919	5.66	255.2		33.4	239.3		2015
2120	5.32	256.1		33.6	246.3		2016
2529	9.65	246.2		32.6	285.3		2017
3053	7.81	252.2		32.7	304.8		2018
3100	7.12	255.3		32.9	320.6		2019
3590	7.56	266.5		34.3	335.8		2020

3-2-1 建制镇市政公用设施水平(2020年)

地区名称 Name of Regions	人口密度 (人/平方公里) Population Density (person/km²)	人均日生 活用水量 (升) Per Capita Daily Water Consumption (liter)	供 水 普及率 (%) Water Coverage Rate (%)	燃 气 普及率 (%) Gas Coverage Rate (%)	人均道 路面积 (平方米) Road Surface Area Per Capita (m²)	排水管 道暗渠 密 度 (公里/平方公里) Density of Drains (km/km²)
全 国 National Total	4249	107.00	89.08	56.94	15.79	7.20
北 京 Beijing	4533	131.59	89.63	64.24	12.50	7.37
天 津 Tianjin	3452	87.52	86.26	75.70	13.67	5.14
河 北 Hebei	3855	87.93	89.66	59.31	11.74	3.87
山 西 Shanxi	3979	83.18	90.12	34.80	15.65	5.50
内 蒙 古 Inner Mongolia	2438	81.88	78.65	28.01	17.77	3.06
辽 宁 Liaoning	3238	118.70	77.40	30.57	15.30	4.08
吉 林 Jilin	3123	87.08	89.38	43.00	14.04	3.64
黑 龙 江 Heilongjiang	3130	83.39	81.81	16.88	16.20	4.49
上 海 Shanghai	5771	127.47	91.89	80.05	8.74	5.14
江 苏 Jiangsu	5122	99.45	98.95	94.63	19.91	10.92
浙 江 Zhejiang	4710	119.15	88.73	59.49	16.48	9.48
安 徽 Anhui	4144	109.43	84.32	46.33	17.62	8.02
福 建 Fujian	4856	121.97	91.17	71.59	17.08	7.87
江 西 Jiangxi	3976	101.50	83.34	43.13	16.62	7.88
山 东 Shandong	4070	85.05	94.19	73.40	16.11	7.23
河 南 Henan	4460	133.71	84.51	31.20	17.29	6.80
湖 北 Hubei	3828	117.20	86.71	46.63	17.14	7.89
湖 南 Hunan	4329	108.56	78.97	39.81	13.46	6.36
广 东 Guangdong	4621	141.97	93.27	81.44	16.43	9.31
广 西 Guangxi	5376	105.98	88.89	76.36	17.55	9.14
海 南 Hainan	3865	98.27	91.54	83.37	16.71	6.92
重 庆 Chongqing	5549	90.88	94.47	74.90	8.35	8.43
四 川 Sichuan	4730	92.40	88.54	69.14	12.26	7.33
贵 州 Guizhou	3759	98.97	89.07	13.04	19.61	7.44
云 南 Yunnan	4977	99.48	94.14	12.66	15.14	9.07
西 藏 Tibet	182	103.70	48.60	9.38	52.96	0.13
陕 西 Shaanxi	4323	80.01	86.52	26.30	14.38	6.75
甘 肃 Gansu	3648	75.81	90.78	11.38	16.54	5.33
青 海 Qinghai	3941	88.66	88.94	28.49	15.90	4.87
宁 夏 Ningxia	3070	91.92	97.26	47.24	17.16	7.36
新 疆 Xinjiang	3069	94.11	89.44	28.25	28.33	4.84

Level of Municipal Public Facilities of Built-up Area of Towns (2020)

污水处理率 (%) Wastewater Treatment Rate (%)	污水处理厂集中处理率 Centralized Treatment Rate of Wastewater Treatment Plants	人均公园绿地面积(平方米) Public Recreational Green Space Per Capita (m²)	绿化覆盖率 (%) Green Coverage Rate (%)	绿地率 (%) Green Space Rate (%)	生活垃圾处理率 (%) Domestic Garbage Treatment Rate (%)	无害化处理率 Domestic Garbage Harmless Treatment Rate	地区名称 Name of Regions
60.98	52.14	2.72	16.88	10.81	89.18	69.55	全　国
66.15	48.74	3.46	23.27	14.36	89.42	84.74	北　京
65.48	60.34	2.00	14.43	8.25	90.45	73.84	天　津
31.33	23.33	1.31	14.11	8.23	79.37	44.03	河　北
30.81	22.39	1.74	19.57	9.98	54.82	15.44	山　西
23.53	19.49	1.65	13.27	7.84	38.21	23.69	内蒙古
48.27	42.46	1.23	14.92	7.35	70.80	30.41	辽　宁
29.79	27.16	2.35	11.37	6.68	98.77	91.71	吉　林
25.95	19.73	1.46	7.96	5.43	43.74	26.47	黑龙江
70.31	67.60	1.80	17.38	11.27	89.33	76.49	上　海
85.29	80.60	7.32	30.17	24.71	99.61	96.59	江　苏
74.56	62.70	2.90	18.79	12.57	92.66	81.83	浙　江
54.12	46.24	1.91	18.80	11.18	97.29	92.30	安　徽
75.53	58.48	5.85	24.76	18.28	98.87	97.85	福　建
35.25	24.38	1.52	12.00	8.51	87.61	51.08	江　西
71.16	55.33	5.08	24.55	16.26	98.78	95.90	山　东
36.17	25.32	1.69	17.92	6.46	77.41	42.18	河　南
50.18	41.28	1.86	16.07	9.05	91.43	68.24	湖　北
45.19	27.53	2.41	22.35	15.10	84.88	48.62	湖　南
61.70	55.87	3.13	14.22	9.36	97.32	90.32	广　东
50.33	38.56	2.72	15.18	9.46	96.95	48.12	广　西
11.70	6.66	0.23	17.63	11.85	96.79	29.89	海　南
85.96	78.33	0.96	11.61	6.96	95.83	73.04	重　庆
63.70	52.69	0.93	7.33	5.27	95.17	76.32	四　川
86.95	82.93	1.39	12.97	7.57	87.15	48.06	贵　州
20.48	17.28	0.80	8.74	5.69	84.68	37.81	云　南
29.86	10.81	0.14	0.30	0.15	95.90	61.40	西　藏
36.59	31.24	1.07	8.16	5.45	66.37	23.04	陕　西
31.90	25.49	0.83	11.89	6.77	63.40	47.25	甘　肃
11.53	7.29	0.34	12.16	9.33	67.46	25.15	青　海
74.87	62.91	2.21	12.61	7.55	96.48	63.85	宁　夏
30.74	17.34	0.82	14.86	10.57	87.45	46.31	新　疆

3-2-2 建制镇基本情况(2020年)

地区名称 Name of Regions	建制镇 个 数 (个) Number of Towns (unit)	建成区 面 积 (公顷) Surface Area of Built- up Districts (hectare)	建 成 区 户籍人口 (万人) Registered Permanent Population (10,000 persons)	建 成 区 常住人口 (万人) Permanant Population (10,000 persons)
全 国 National Total	18822	4338502.52	16596.22	18433.28
北 京 Beijing	115	29929.05	75.24	135.68
天 津 Tianjin	113	46147.13	116.89	159.32
河 北 Hebei	978	170861.25	603.91	658.72
山 西 Shanxi	482	66965.61	256.79	266.46
内 蒙 古 Inner Mongolia	435	97779.52	239.96	238.38
辽 宁 Liaoning	612	95844.56	290.94	310.33
吉 林 Jilin	391	78831.46	255.59	246.18
黑 龙 江 Heilongjiang	471	87344.21	287.72	273.42
上 海 Shanghai	99	133982.75	333.16	773.27
江 苏 Jiangsu	667	272939.27	1220.33	1398.00
浙 江 Zhejiang	571	212508.38	707.49	1000.85
安 徽 Anhui	883	250475.71	987.76	1038.02
福 建 Fujian	557	142562.88	623.94	692.27
江 西 Jiangxi	726	147676.52	600.68	587.11
山 东 Shandong	1071	399975.75	1499.23	1627.95
河 南 Henan	1068	288045.45	1271.06	1284.67
湖 北 Hubei	699	223756.91	859.01	856.49
湖 南 Hunan	$10^4 1$	253152.01	1074.21	1096.02
广 东 Guangdong	1002	348814.20	1246.33	1611.70
广 西 Guangxi	702	98649.90	565.72	530.32
海 南 Hainan	158	27536.28	101.06	106.43
重 庆 Chongqing	584	78467.67	426.76	435.45
四 川 Sichuan	1766	241670.98	1006.02	1143.22
贵 州 Guizhou	773	140346.89	547.49	527.56
云 南 Yunnan	592	78861.96	380.90	392.51
西 藏 Tibet	73	56895.74	7.86	10.36
陕 西 Shaanxi	927	127485.47	546.72	551.08
甘 肃 Gansu	783	68221.51	240.50	248.84
青 海 Qinghai	106	10674.51	43.88	42.07
宁 夏 Ningxia	76	18961.99	55.38	58.22
新 疆 Xinjiang	301	43137.00	123.70	132.37

Summary of Towns(2020)

设有村镇建设管理机构的个数 (个) Number of Towns with Construction Management Institution (unit)	村镇建设管理人员 (人) Number of Construction Management Personnel (person)	专职人员 Full-time Staff	有总体规划的建制镇个数 (个) Number of Towns with Master Plans (unit)	本年编制 Compiled This Year	本年规划编制投入 (万元) Input in Planning this Year (10,000 RMB)	地区名称 Name of Regions
17413	90505	57075	16833	1039	383162.44	全 国
106	1290	569	98	3	12011.70	北 京
112	701	405	83	9	2701.43	天 津
879	3483	2210	732	46	3797.40	河 北
341	640	349	349	7	3130.50	山 西
396	1386	919	394	18	819.00	内 蒙 古
598	1646	1092	529	21	975.65	辽 宁
384	1252	829	279	14	782.60	吉 林
453	1052	652	384	33	520.02	黑 龙 江
98	1165	714	82	5	3904.82	上 海
663	7485	5128	655	50	23366.69	江 苏
546	5326	3338	545	39	21346.29	浙 江
814	3873	2560	823	65	14161.71	安 徽
545	2005	1317	524	12	12035.95	福 建
707	2959	1812	706	52	22549.57	江 西
1070	7690	5027	1021	77	25937.39	山 东
10⁴5	7300	4584	943	49	21540.92	河 南
669	4576	2840	655	33	12916.40	湖 北
950	5489	3349	961	73	18712.37	湖 南
950	9096	5269	908	56	56068.44	广 东
697	3874	2475	685	21	5111.42	广 西
155	429	260	154	1	5181.66	海 南
581	2491	1761	557	59	3855.84	重 庆
1495	4836	3207	1456	71	44037.83	四 川
763	2568	1775	727	87	17973.78	贵 州
584	2738	1687	564	21	12618.55	云 南
12	47	9	47	7	2468.78	西 藏
810	2391	1347	821	60	16248.63	陕 西
602	1699	993	693	33	13318.10	甘 肃
82	128	71	99	6	176.00	青 海
68	216	113	70	5	269.10	宁 夏
238	674	414	289	6	4623.90	新 疆

3-2-3 建制镇供水(2020年)

地区名称 Name of Regions	集中供水的 建制镇个数 （个） Number of Towns with Access to Piped Water (unit)	占全部建制 镇的比例 （%） Percentage of Total Rate (%)	公共供水综合 生 产 能 力 （万立方米／日） Integrated Production Capacity of Public Water Supply Facilities (10,000 m³/day)	自备水源单位 综合生产能力 （万立方米／日） Integrated Production Capacity of Self-built Water Supply Facilities (10,000 m³/day)	年供水 总 量 （万立方米） Annual Supply of Water (10,000 m³)
全 国 National Total	18285	97.15	9753.15	2324.67	1452136.81
北 京 Beijing	114	99.13	68.01	58.10	11319.37
天 津 Tianjin	110	97.35	70.29	10.16	11351.03
河 北 Hebei	914	93.46	174.33	123.90	41400.8
山 西 Shanxi	465	96.47	65.51	21.24	17493.38
内 蒙 古 Inner Mongolia	428	98.39	61.31	22.48	13358.58
辽 宁 Liaoning	513	83.82	125.61	40.57	18705.52
吉 林 Jilin	386	98.72	88.34	27.27	12708.65
黑 龙 江 Heilongjiang	455	96.60	54.95	14.84	9609.7
上 海 Shanghai	97	97.98	808.16	61.62	101885.81
江 苏 Jiangsu	664	99.55	1085.09	75.43	123853.82
浙 江 Zhejiang	555	97.20	733.22	141.42	114143.22
安 徽 Anhui	873	98.87	468.93	90.67	70013.5
福 建 Fujian	549	98.56	358.47	73.70	73856.2
江 西 Jiangxi	711	97.93	222.30	45.00	36326.5
山 东 Shandong	1067	99.63	615.65	235.82	126469.5
河 南 Henan	1045	97.85	836.54	259.64	110119.6
湖 北 Hubei	672	96.14	483.82	92.41	63562.19
湖 南 Hunan	982	94.33	586.74	128.34	64786.89
广 东 Guangdong	977	97.50	1308.37	334.21	206824.24
广 西 Guangxi	695	99.00	179.80	55.22	31219.3
海 南 Hainan	158	100.00	27.45	6.94	5652.64
重 庆 Chongqing	579	99.14	126.17	19.02	22591.56
四 川 Sichuan	1756	99.43	500.20	131.40	58480.87
贵 州 Guizhou	745	96.38	269.78	97.78	29812.91
云 南 Yunnan	591	99.83	141.72	76.88	26082.43
西 藏 Tibet	38	52.05	1.29	0.95	306.2
陕 西 Shaanxi	913	98.49	82.54	37.42	23441.62
甘 肃 Gansu	762	97.32	76.34	15.68	10915.06
青 海 Qinghai	97	91.51	44.16	7.44	2688.83
宁 夏 Ningxia	76	100.00	18.96	4.33	3454.57
新 疆 Xinjiang	298	99.00	69.09	14.79	9702.32

Water Supply of Towns(2020)

年生活 用水量 Annual Domestic Water Consumption	年生产 用水量 Annual Water Consumption for Production	供水管道 长　度 （公里） Length of Water Supply Pipelines (km)	本年新增 Added This Year	用水人口 （万人） Population with Access to Water (10,000 persons)	地区名称 Name of Regions
641351.94	677517.22	624464.93	33861.83	16421.02	全　　国
5841.39	3970.23	4882.26	47.22	121.62	北　京
4390.33	6541.18	4586.54	207.81	137.44	天　津
18954.54	19482.71	12996.22	582.70	590.59	河　北
7290.74	8526.79	9992.86	394.71	240.14	山　西
5603.34	6330.66	7529.41	97.95	187.49	内　蒙　古
10406.33	7172.48	9218.35	298.79	240.19	辽　宁
6993.59	4565.40	7630.17	412.90	220.04	吉　林
6808.12	2349.41	8905.28	166.41	223.68	黑　龙　江
33059.28	59540.44	10187.09	119.44	710.55	上　海
50212.87	67785.87	58581.09	1759.55	1383.35	江　苏
38620.17	65626.80	36978.66	1517.77	888.01	浙　江
34957.90	25723.96	39177.18	2911.88	875.24	安　徽
28096.38	38894.79	18084.94	1181.08	631.11	福　建
18127.50	14605.62	17403.12	1461.88	489.29	江　西
47600.03	69646.21	37108.77	2496.55	1533.30	山　东
52982.57	46982.37	44032.38	2937.87	1085.65	河　南
31769.64	24504.11	36377.02	1945.57	742.64	湖　北
34295.63	24985.48	40805.65	1795.00	865.53	湖　南
77894.72	99570.66	55018.89	2348.75	1503.21	广　东
18235.78	10052.66	11967.89	688.40	471.41	广　西
3494.50	2033.00	5374.58	259.37	97.43	海　南
13644.67	7522.50	8136.14	600.35	411.35	重　庆
34140.08	20531.68	44436.71	3175.66	1012.22	四　川
16973.05	9753.05	26440.48	2405.59	469.88	贵　州
13416.74	10821.17	21704.94	1285.10	369.50	云　南
190.62	80.13	2980.59	68.01	5.04	西　藏
13923.52	8550.91	15397.11	1256.76	476.80	陕　西
6250.45	4042.87	12092.07	607.12	225.89	甘　肃
1210.73	1287.86	1187.90	29.83	37.41	青　海
1899.74	1295.25	2519.30	152.18	56.63	宁　夏
4066.99	4740.97	12731.34	649.63	118.39	新　疆

3-2-4 建制镇燃气、供热、道路桥梁(2020年)

地区名称 Name of Regions	用气人口 (万人) Population with Access to Gas (10,000 persons)	集中供热面积 (万平方米) Area of Centrally Heated District (10,000 m²)	道路长度 (公里) Length of Roads (km)	本年新增 Added This Year	本年更新改造 Renewal and Upgrading during the Reported Year
全 国 **National Total**	**10495.68**	**44363.41**	**438981.12**	**26651.29**	**25506.18**
北 京 Beijing	87.16	1998.52	2728.25	39.03	76.51
天 津 Tianjin	120.61	4425.32	3078.27	35.64	130.21
河 北 Hebei	390.66	3327.52	13765.99	220.85	937.19
山 西 Shanxi	92.72	2006.70	7264.62	258.44	346.06
内 蒙 古 Inner Mongolia	66.77	2689.61	6904.91	80.92	212.42
辽 宁 Liaoning	94.88	3740.43	8434.65	184.97	429.28
吉 林 Jilin	105.85	3274.94	5746.96	42.14	437.85
黑 龙 江 Heilongjiang	46.16	2410.82	7835.43	22.47	169.38
上 海 Shanghai	618.98	56.42	6142.12	59.86	179.78
江 苏 Jiangsu	1322.99	185.68	38095.46	1198.86	1680.08
浙 江 Zhejiang	595.43	237.43	21496.02	792.93	1073.76
安 徽 Anhui	480.94		27948.69	2616.02	1453.65
福 建 Fujian	495.62	26.69	16692.88	792.97	645.75
江 西 Jiangxi	253.24	318.82	15549.12	1329.04	1122.54
山 东 Shandong	1194.84	13224.32	37005.07	2430.94	3292.04
河 南 Henan	400.84	957.73	35736.23	3640.76	1832.21
湖 北 Hubei	399.39	300.60	22536.87	1796.79	1658.70
湖 南 Hunan	436.30	591.80	24514.99	1512.27	1502.52
广 东 Guangdong	1312.50		37048.61	1924.07	1938.53
广 西 Guangxi	404.95	53.60	13298.14	737.95	720.53
海 南 Hainan	88.73		2993.07	156.03	135.01
重 庆 Chongqing	326.16	13.41	5334.79	353.14	394.24
四 川 Sichuan	790.40		23109.89	2240.28	1345.79
贵 州 Guizhou	68.77	85.63	16229.67	1441.58	1616.77
云 南 Yunnan	49.68	21.53	10130.50	932.47	844.81
西 藏 Tibet	0.97	8.72	459.35	27.61	3.63
陕 西 Shaanxi	144.91	1043.18	13201.12	932.53	610.74
甘 肃 Gansu	28.33	1192.90	6750.20	340.94	292.45
青 海 Qinghai	11.99	293.95	1126.51	94.83	33.75
宁 夏 Ningxia	27.50	773.25	1505.48	74.30	78.94
新 疆 Xinjiang	37.40	1103.89	6317.26	340.66	311.06

Gas, Central Heating, Road and Bridges of Towns(2020)

安装路灯的道路长度 Roads with Lamps	道路面积 （万平方米） Surface Area of Roads (10,000 m²)	本年新增 Added This year	本年更新改造 Renewal and Upgrading during the Reported Year	桥梁座数 （座） Number of Bridges (unit)	本年新增 Added This year	地区名称 Name of Regions
139247.26	291002.00	18444.70	15818.19	80618	3719	全　国
1575.69	1696.68	26.23	40.37	327	3	北　京
1466.81	2178.11	28.33	87.85	514	16	天　津
4916.03	7733.04	216.27	508.51	1840	62	河　北
1377.32	4170.31	208.13	161.34	720	22	山　西
1727.09	4236.94	53.97	138.55	582	16	内　蒙古
2279.74	4748.48	165.38	269.54	1931	64	辽　宁
1667.72	3456.41	47.87	250.89	922	24	吉　林
1756.32	4429.80	14.41	92.33	556	9	黑　龙江
2393.72	6760.14	90.85	170.97	4226	59	上　海
17736.85	27830.78	997.79	1239.60	7483	239	江　苏
8443.02	16497.96	732.34	887.86	10894	159	浙　江
8854.62	18285.91	1679.34	711.69	5644	479	安　徽
6544.37	11820.95	781.85	504.20	2669	121	福　建
3330.60	9755.28	856.51	721.29	2862	202	江　西
13921.13	26233.82	1704.18	2067.86	6573	487	山　东
8510.63	22214.52	2187.74	1246.35	4082	367	河　南
5508.41	14683.41	1131.09	1249.12	4367	186	湖　北
6112.61	14756.11	954.28	757.94	3180	131	湖　南
14689.67	26479.57	1451.20	1316.76	6334	181	广　东
3809.19	9305.06	529.62	461.25	1694	122	广　西
1229.59	1778.35	152.90	78.31	477	13	海　南
2391.45	3637.59	300.30	291.36	1638	33	重　庆
6040.66	14018.23	1479.33	692.96	4055	233	四　川
3604.89	10345.55	967.91	733.89	1401	136	贵　州
2274.24	5941.58	593.51	380.76	1217	78	云　南
94.58	548.72	17.29	5.15	89	13	西　藏
3165.63	7924.54	516.65	353.07	2139	137	陕　西
1766.57	4116.36	288.58	220.57	1168	62	甘　肃
189.53	668.66	85.34	23.83	171	8	青　海
481.33	999.16	36.17	37.64	128		宁　夏
1387.25	3749.98	149.34	116.38	735	57	新　疆

3-2-5　建制镇排水和污水处理(2020年)
Drainage and Wastewater Treatment of Towns(2020)

地区名称 Name of Regions	对生活污水进行处理的建制镇 Town with Domestic Wastewater Treated		污水处理厂 Wastewater Treatment Plant		污水处理装置处理能力(万立方米/日) Treatment Capacity of Wastewater Treatment Facilities (10,000 m³/day)	排水管道长度(公里) Length of Drainage Pipelines (km)		排水暗渠长度(公里) Length of Drains (km)	
	个数(个) Number (unit)	比例(%) Rate (%)	个数(个) Number of Wastewater Treatment Plants (unit)	处理能力(万立方米/日) Treatment Capacity (10,000 m³/day)			本年新增 Added This Year		本年新增 Added This Year
全　国 National Total	12300	65.35	11374	2740.05	2157.36	198409.18	18611.32	113879.37	8706.30
北　京 Beijing	97	84.35	113	39.67	40.82	1380.91	36.28	824.64	4.63
天　津 Tianjin	94	83.19	154	19.99	7.98	2001.66	337.59	369.60	19.36
河　北 Hebei	258	26.38	136	86.40	67.08	4281.91	421.58	2329.05	105.86
山　西 Shanxi	224	46.47	139	31.87	19.28	2200.95	282.41	1484.01	80.89
内　蒙　古 Inner Mongolia	96	22.07	69	12.70	10.56	2186.09	223.71	808.87	66.93
辽　宁 Liaoning	197	32.19	142	33.67	21.83	2427.51	80.78	1481.15	43.22
吉　林 Jilin	138	35.29	88	32.30	32.89	1993.23	270.13	872.70	93.90
黑　龙　江 Heilongjiang	115	24.42	85	14.16	11.11	2158.07	280.12	1764.36	93.30
上　海 Shanghai	95	95.96	18	37.49	30.11	5901.98	237.93	982.29	9.31
江　苏 Jiangsu	665	99.70	700	378.62	202.99	21129.52	1067.47	8674.15	464.39
浙　江 Zhejiang	536	93.87	281	159.77	156.51	15193.45	873.53	4957.98	211.16
安　徽 Anhui	749	84.82	685	91.52	75.51	12183.43	1224.89	7903.41	807.38
福　建 Fujian	548	98.38	540	129.01	221.75	7553.54	674.66	3663.74	380.76
江　西 Jiangxi	470	64.74	417	24.78	21.75	6375.42	721.95	5267.99	453.44
山　东 Shandong	1030	96.17	921	297.20	257.73	17509.65	909.29	11422.86	591.68
河　南 Henan	474	44.38	376	108.27	86.73	11493.68	1358.05	8095.46	792.17
湖　北 Hubei	643	91.99	541	144.53	87.25	11900.84	1180.09	5760.29	565.40
湖　南 Hunan	707	67.92	464	83.62	73.54	10246.28	934.41	5851.03	336.57
广　东 Guangdong	779	77.74	831	537.20	400.56	18304.95	2899.03	14165.40	1491.95
广　西 Guangxi	523	74.50	464	46.48	41.90	5114.07	343.85	3906.96	339.68
海　南 Hainan	16	10.13	15	5.01	1.55	1125.19	150.39	779.82	173.29
重　庆 Chongqing	574	98.29	724	78.91	38.33	4473.04	662.33	2140.18	126.44
四　川 Sichuan	1379	78.09	1959	195.27	138.19	11114.60	1406.80	6608.94	506.86
贵　州 Guizhou	591	76.46	492	51.49	45.75	5952.26	743.69	4495.84	319.32
云　南 Yunnan	331	55.91	261	29.95	14.53	3597.53	357.92	3554.07	269.03
西　藏 Tibet	12	16.44	6	0.14	0.01	44.42	3.57	29.27	0.41
陕　西 Shaanxi	421	45.42	388	31.44	28.90	5156.43	479.10	3454.50	228.11
甘　肃 Gansu	309	39.46	206	17.41	7.08	2451.78	203.85	1184.96	91.37
青　海 Qinghai	31	29.25	7	1.69	4.51	423.22	5.34	96.46	0.59
宁　夏 Ningxia	70	92.11	81	7.63	1.11	963.02	56.53	432.75	18.80
新　疆 Xinjiang	128	42.52	71	11.85	9.50	1570.55	184.05	516.64	20.10

3-2-6 建制镇园林绿化及环境卫生(2020年)
Landscaping and Environmental Sanitation of Towns(2020)

地区名称 Name of Regions	园林绿化(公顷) Landscaping(hectare)				环境卫生 Environmental Sanitation		
	绿化覆盖面积 Green Coverage Area	绿地面积 Area of Parks and Green Space	本年新增 Added This Year	公园绿地面积 Public Green Space	生活垃圾中转站 (座) Number of Garbage Transfer Station (unit)	环卫专用车辆设备 (辆) Number of Special Vehicles for Environmental Sanitation (unit)	公共厕所 (座) Number of Latrines (unit)
全 国 National Total	732157.16	468789.11	22973.78	50093.27	28737	119983	135070
北 京 Beijing	6964.98	4299.19	383.29	468.86	288	895	1182
天 津 Tianjin	6660.93	3805.51	126.05	319.08	241	1376	1360
河 北 Hebei	24113.06	14061.58	1211.35	864.86	1176	4144	3730
山 西 Shanxi	13103.79	6683.85	232.15	463.18	371	2897	2520
内 蒙 古 Inner Mongolia	12977.57	7668.18	237.03	392.54	496	1824	3023
辽 宁 Liaoning	14304.08	7043.87	261.88	382.04	497	3334	2225
吉 林 Jilin	8964.69	5264.68	782.83	578.59	414	1999	1260
黑 龙 江 Heilongjiang	6954.94	4740.58	419.68	399.82	316	1997	1405
上 海 Shanghai	23290.68	15101.82	588.66	1390.58	298	2857	2306
江 苏 Jiangsu	82339.75	67449.89	1963.75	10237.56	1238	7824	7501
浙 江 Zhejiang	39923.71	26722.33	1461.47	2906.17	1302	8277	15609
安 徽 Anhui	47084.32	27993.23	1496.60	1977.70	1519	8820	7611
福 建 Fujian	35299.22	26058.30	670.09	4049.27	816	4109	5927
江 西 Jiangxi	17715.12	12568.00	891.98	891.16	1193	2780	4229
山 东 Shandong	98180.03	65034.22	2282.10	8266.47	1329	6952	7714
河 南 Henan	51616.81	18621.84	1823.17	2175.47	2425	9044	6949
湖 北 Hubei	35964.84	20248.77	1190.23	1592.05	1260	5299	10049
湖 南 Hunan	56584.23	38233.17	973.47	2640.81	1837	4894	6487
广 东 Guangdong	49585.27	32651.68	1286.39	5041.15	2240	11768	11969
广 西 Guangxi	14977.36	9331.00	986.40	1441.99	997	4624	2380
海 南 Hainan	4854.17	3264.42	113.36	24.96	145	735	447
重 庆 Chongqing	9108.20	5463.57	297.53	419.49	645	2048	2784
四 川 Sichuan	17724.35	12731.29	734.36	1068.52	3334	6559	8542
贵 州 Guizhou	18198.76	10622.28	555.22	735.36	1017	3934	4382
云 南 Yunnan	6890.66	4487.39	339.24	314.68	583	2325	3932
西 藏 Tibet	168.37	87.77	5.71	1.47	61	218	357
陕 西 Shaanxi	10398.91	6944.95	438.50	590.36	1147	3632	4110
甘 肃 Gansu	8110.11	4619.43	654.12	207.38	1098	2717	3025
青 海 Qinghai	1297.94	995.44	116.40	14.27	83	401	517
宁 夏 Ningxia	2391.04	1431.85	98.16	128.82	131	494	370
新 疆 Xinjiang	6409.27	4559.03	352.61	108.61	240	1206	1168

3-2-7 建制镇房屋(2020年)

地区名称 Name of Regions	住宅 Residential Building						公共建筑	
	年末实有建筑面积(万平方米) Total Floor Space of Buildings (year-end) (10,000 m²)	混合结构以上 Mixed Strucure and Above	本年竣工建筑面积(万平方米) Floor Space Completed This Year (10,000 m²)	混合结构以上 Mixed Strucure and above	房地产开发 Real Estate Development	人均住宅建筑面积(平方米) Per Capita Floor Space (m²)	年末实有建筑面积(万平方米) Total Floor Space of Buildings (year-end) (10,000 m²)	混合结构以上 Mixed Strucure and Above
全 国 National Total	614392.01	496397.65	27911.43	21470.78	11255.07	37.02	151092.01	137562.61
北 京 Beijing	4236.13	3666.20	181.17	92.08	156.68	56.30	1757.27	1555.29
天 津 Tianjin	5744.22	5086.90	278.19	198.88	213.51	49.14	1197.77	1170.66
河 北 Hebei	18989.16	15360.29	684.32	542.20	276.45	31.44	4243.40	3988.48
山 西 Shanxi	8103.70	4817.22	276.53	129.63	36.81	31.56	1430.67	1088.27
内 蒙 古 Inner Mongolia	7008.91	5856.40	117.05	52.68	38.49	29.21	2471.93	2368.63
辽 宁 Liaoning	8438.34	6168.74	346.52	310.22	105.50	29.00	2776.86	2439.05
吉 林 Jilin	7736.20	6098.78	261.08	251.64	246.65	30.27	1913.89	1828.96
黑 龙 江 Heilongjiang	7425.02	6536.38	291.42	27.12	19.20	25.81	1808.94	1708.59
上 海 Shanghai	23449.54	21698.30	1239.79	665.72	1029.36	70.39	10835.16	10686.28
江 苏 Jiangsu	56559.80	51499.10	2655.41	2394.02	1360.76	46.35	14524.54	13547.97
浙 江 Zhejiang	38437.88	32608.05	2229.78	1585.30	990.70	54.33	8453.31	7894.07
安 徽 Anhui	32853.84	25786.41	1824.82	1310.78	695.96	33.26	6473.11	5793.83
福 建 Fujian	25416.57	20754.22	793.08	646.82	371.19	40.74	5573.00	5106.91
江 西 Jiangxi	23577.22	17929.68	702.42	625.79	156.44	39.25	4193.07	3860.62
山 东 Shandong	51410.17	39436.97	2978.16	2460.55	1133.23	34.29	15374.74	14462.56
河 南 Henan	44093.90	34203.40	2332.80	1772.66	443.77	34.69	7772.44	7075.78
湖 北 Hubei	28181.07	22608.12	1008.91	851.37	251.65	32.81	5825.72	5040.44
湖 南 Hunan	35125.18	29020.96	1410.92	1225.27	244.70	32.70	7920.70	6312.94
广 东 Guangdong	46681.96	38102.42	3095.83	1997.76	2143.84	37.46	9941.74	8813.62
广 西 Guangxi	18511.92	14938.86	678.03	619.33	163.21	32.72	4646.36	4320.19
海 南 Hainan	4051.16	4023.62	304.82	293.49	234.77	40.09	699.45	681.43
重 庆 Chongqing	15768.00	13825.50	225.91	210.55	74.69	36.95	2503.90	2369.33
四 川 Sichuan	36430.93	29761.55	1405.52	1028.21	342.17	36.21	7495.79	6908.35
贵 州 Guizhou	21823.50	13317.03	970.34	802.37	203.07	39.86	3497.21	2806.48
云 南 Yunnan	12862.06	9434.75	547.55	455.28	87.03	33.77	2757.66	2470.76
西 藏 Tibet	796.99	614.30	6.77	2.95	5.55	101.39	74.33	59.32
陕 西 Shaanxi	15657.56	12261.91	433.40	390.59	88.81	28.64	3308.08	2959.99
甘 肃 Gansu	7330.95	5344.38	277.43	209.74	113.83	30.48	8212.31	7932.54
青 海 Qinghai	1341.63	994.40	32.29	27.12	2.30	30.57	621.41	564.40
宁 夏 Ningxia	2304.13	1906.62	51.71	39.64	19.67	41.61	710.13	676.25
新 疆 Xinjiang	4044.38	2736.20	269.48	251.02	5.06	32.70	2077.10	1070.61

Building Construction of Towns(2020)

Public Building			生产性建筑　Industrial Building					
本年竣工建筑面积(万平方米) Floor Space Completed This Year (10,000 m²)	混合结构以上 Mixed Strucure and Above	房地产开发 Real Estate Development	年末实有建筑面积(万平方米) Total Floor Space of Buildings (year-end) (10,000m²)	混合结构以上 Mixed Strucure and Above	本年竣工建筑面积(万平方米) Floor Space Completed This Year (10,000 m²)	混合结构以上 Mixed Strucure and Above	房地产开发 Real Estate Development	地区名称 Name of Regions
7328.32	5681.46	880.54	216585.21	188223.30	12588.91	9306.59	669.31	全　国
18.33	18.03	15.00	1314.43	1120.87	33.98	33.98	0.00	北　京
29.72	25.01	14.14	4041.31	3865.84	54.07	53.16	0.00	天　津
245.30	191.09	129.57	7018.24	5894.44	217.06	176.40	19.96	河　北
46.74	44.46	0.90	1974.95	1437.76	55.86	54.36	0.00	山　西
25.04	19.31	0.91	1579.06	1397.51	31.91	28.04	1.50	内 蒙 古
21.11	17.45	6.44	2711.28	2060.45	27.28	24.46	0.31	辽　宁
44.46	43.70	0.02	1804.66	1569.61	42.29	34.06	0.00	吉　林
48.10	4.92	0.00	1821.70	1707.34	6.89	6.43	0.00	黑 龙 江
137.33	132.73	15.86	13443.33	13102.83	314.37	303.36	23.42	上　海
458.42	415.83	68.43	35889.69	33605.13	1803.80	1677.09	71.20	江　苏
508.38	406.38	143.32	32775.70	29961.21	1740.84	1650.54	149.85	浙　江
320.68	272.60	32.84	7937.26	6093.15	461.62	372.74	36.35	安　徽
179.94	164.37	4.92	8705.86	7370.76	267.95	243.73	30.56	福　建
345.16	328.13	19.41	5014.95	4689.96	374.28	282.55	11.23	江　西
667.93	596.66	73.12	31924.72	28500.22	1517.54	1373.27	51.02	山　东
640.48	585.05	53.22	7577.90	6446.55	518.21	444.63	58.31	河　南
435.69	324.39	83.47	5797.48	4566.72	430.98	364.20	40.58	湖　北
405.03	353.47	15.00	4327.28	3839.85	368.80	354.48	13.16	湖　南
462.07	392.58	82.27	16834.90	13098.17	905.31	819.87	84.35	广　东
118.51	98.55	1.90	2492.03	1979.32	162.25	135.14	6.51	广　西
14.77	12.30	1.03	312.12	306.47	2.32	2.32	0.15	海　南
57.09	54.92	0.60	2743.29	1815.80	50.26	48.49	0.25	重　庆
653.50	599.48	43.01	5737.72	4957.07	242.73	158.89	33.91	四　川
171.24	149.86	14.30	4864.27	2237.44	2365.28	111.10	13.40	贵　州
128.15	118.82	1.77	2108.47	1648.66	76.35	65.63	1.10	云　南
5.04	2.37	0.20	57.19	44.91	13.85	12.20	0.70	西　藏
124.18	114.40	4.44	1915.30	1653.49	132.52	122.85	3.72	陕　西
78.08	64.15	6.36	1992.74	1651.31	276.42	265.21	1.00	甘　肃
19.11	18.90	1.10	370.93	316.73	2.07	1.62	0.15	青　海
65.68	42.10	24.41	503.46	480.52	13.54	12.99	0.00	宁　夏
853.05	69.44	22.56	992.99	803.21	78.29	72.77	16.64	新　疆

3-2-8 建制镇建设投入(2020年)

计量单位: 万元

地区名称 Name of Regions	本年建设投入合计 Total Construction Input of This Year	房屋　Building					小　计 Input of Municipal Public Facilities	供　水 Water Supply
		小　计 Input of House	房地产开发 Real Estate Development	住　宅 Residential Building	公共建筑 Public Building	生产性建筑 Industrial Building		
全　国　National Total	**96782066**	**76306857**	**33023322**	**50237848**	**11261295**	**14807714**	**20475209**	**1704760**
北　京　Beijing	775859	570069	88280	498059	65232	6778	205790	18318
天　津　Tianjin	2655359	2276146	1601388	1940368	172346	163432	379213	16703
河　北　Hebei	1841883	1553045	339180	837736	365009	350300	288839	35024
山　西　Shanxi	646306	437896	76190	270920	109400	57577	208409	9564
内　蒙　古　Inner Mongolia	344340	194324	39937	118287	34806	41231	150016	17888
辽　宁　Liaoning	341077	206913	120313	152516	24785	29611	134164	11510
吉　林　Jilin	1108445	906778	774901	803755	52943	50080	201667	16761
黑　龙　江　Heilongjiang	670155	511801	34039	432690	67788	11323	158354	10499
上　海　Shanghai	9268980	8363005	4914790	5860379	1547799	954827	905975	37002
江　苏　Jiangsu	10610195	8642937	2953206	5043929	824934	2774074	1967257	143867
浙　江　Zhejiang	11932486	10016184	3872760	6294496	969728	2751960	1916302	113096
安　徽　Anhui	5579674	4283142	2054005	3059763	515104	708276	1296531	106899
福　建　Fujian	2476900	1812943	1011776	1278138	223730	311075	663957	75126
江　西　Jiangxi	1725633	1116840	244770	645006	205970	265864	608793	65503
山　东　Shandong	9965465	7863203	2022352	4524202	1054625	2284376	2102262	185466
河　南　Henan	3953117	3109658	1259821	2104938	411304	593416	843459	68390
湖　北　Hubei	2450023	1583167	546454	934076	325013	324078	866856	69821
湖　南　Hunan	2437349	1700602	392843	1192367	297332	210903	736747	107419
广　东　Guangdong	12970906	10795592	7394220	7033470	2060853	1701269	2175314	144152
广　西　Guangxi	1640172	1141900	230668	854227	116459	171214	498272	55726
海　南　Hainan	315618	147855	45340	125232	18922	3702	167763	9355
重　庆　Chongqing	930760	481523	141773	310172	98822	72529	449237	24270
四　川　Sichuan	3924835	2616779	693991	1712354	578095	326329	1308056	103920
贵　州　Guizhou	2870405	2096114	1001647	1696871	224367	174876	774291	98684
云　南　Yunnan	1500150	1161047	308865	874711	189793	96543	339103	44472
西　藏　Tibet	213797	194312	21006	164690	25643	3980	19485	5706
陕　西　Shaanxi	1575220	895695	180789	512113	160976	222607	679525	72204
甘　肃　Gansu	921364	688523	314337	469473	139414	79636	232840	18695
青　海　Qinghai	143036	89784	8788	49402	37135	3246	53253	1377
宁　夏　Ningxia	287887	234822	175874	99844	120100	14877	53065	3223
新　疆　Xinjiang	704670	614258	159019	343664	222868	47726	90412	14123

Construction Input of Towns(2020)

Measurement Unit: 10,000 RMB

市政公用设施 Municipal Public Facilities					园林绿化	环境卫生		其 他	地区名称
燃 气 Gas Supply	集中供热 Central Heating	道路桥梁 Road and Bridge	排 水 Drainage	污水处理 Wastewater Treatment	园林绿化 Landscaping	环境卫生 Environ- mental Sanitation	垃圾处理 Garbage Treatment	Other	Name of Regions
889131	**576376**	**6681606**	**5090016**	**3740511**	**1771294**	**2259535**	**1219604**	**1502491**	全　　国
20789	26106	76005	12660	8174	21751	20248	10799	9912	北　　京
7534	23316	110546	37623	18292	41493	14932	5494	127067	天　　津
62519	28004	47787	60222	39299	17886	33936	20051	3462	河　　北
17828	64681	23122	48312	38760	9849	29107	7094	5946	山　　西
573	27551	36405	35106	23670	12160	15464	8427	4868	内　蒙　古
1541	33666	24031	31791	22517	6263	18399	11249	6964	辽　　宁
3975	15982	80913	55074	42874	8799	14790	8064	5373	吉　　林
424	10572	10988	93143	65943	4849	16767	10569	11113	黑　龙　江
22290	85	319631	172666	111305	87913	136016	56480	130374	上　　海
116500	12972	656994	397803	254302	262606	254069	111398	122446	江　　苏
47990	8043	669172	460605	307234	249697	219359	98606	148341	浙　　江
54432	450	421917	325208	219698	152175	146997	73668	88454	安　　徽
12161	20	239901	156341	110846	54184	90094	58957	36130	福　　建
12965	1269	236212	126511	75082	52882	56456	30465	56995	江　　西
177123	188609	633711	292037	187414	255302	228710	109604	141304	山　　东
65083	31633	336774	134364	83173	89445	90094	40537	27675	河　　南
25857	11692	210735	363033	292091	50836	67229	42240	67652	湖　　北
30323	280	157506	247295	188959	52381	84846	50924	56698	湖　　南
32778	5	578294	994559	886441	96349	220464	113539	108712	广　　东
3433	120	228095	109711	70378	25294	40139	28551	35755	广　　西
2759	304	34940	40732	35491	3530	74333	68806	1811	海　　南
18727		155400	103630	79949	29263	102387	86949	15561	重　　庆
90918		513530	367419	306873	62527	89965	54359	79777	四　　川
7543	223	397190	161661	130023	19403	54935	38037	34652	贵　　州
2752		168269	55392	33772	13253	35343	23064	19623	云　　南
0	430	6708	3099	1372	263	1625	1066	1654	西　　藏
40720	15266	190980	108848	42525	66315	54613	22954	130579	陕　　西
4469	50034	62030	50998	42013	11440	23782	15132	11391	甘　　肃
270	8532	28180	9254	3760	2005	3137	1759	499	青　　海
2133	7660	11030	12052	8029	5313	9665	1977	1988	宁　　夏
2723	8871	14611	22868	10251	5865	11634	8786	9718	新　　疆

3-2-9 乡市政公用设施水平(2020年)

地区名称 Name of Regions	人口密度 (人/平方公里) Population Density (person/km²)	人均日生 活用水量 (升) Per Capita Daily Water Consumption (liter)	供 水 普及率 (%) Water Coverage Rate (%)	燃 气 普及率 (%) Gas Coverage Rate (%)	人均道路 面 积 (平方米) Road Surface Area Per Capita (m²)	排水管道 暗渠密度 (公里/平方公里) Density of Drains (km/km²)
全 国 National Total	**3718**	**97.02**	**83.85**	**30.85**	**21.41**	**7.18**
北 京 Beijing	4051	129.20	92.31	39.34	16.62	7.10
天 津 Tianjin	835	79.70	79.18	65.67	17.70	2.28
河 北 Hebei	3191	87.41	84.17	45.54	15.77	3.33
山 西 Shanxi	3011	81.62	77.87	19.41	18.03	4.09
内 蒙 古 Inner Mongolia	2010	84.56	68.51	19.51	24.68	3.26
辽 宁 Liaoning	3483	95.83	53.75	14.66	23.70	3.72
吉 林 Jilin	2815	83.70	85.26	42.79	20.53	4.15
黑 龙 江 Heilongjiang	2382	81.09	88.41	8.25	23.77	3.57
上 海 Shanghai	3454	113.67	99.79	16.20	27.90	15.40
江 苏 Jiangsu	5470	90.07	99.33	96.85	23.28	15.86
浙 江 Zhejiang	3638	109.01	91.60	49.79	27.00	13.65
安 徽 Anhui	3811	102.27	87.39	36.70	23.10	10.51
福 建 Fujian	4970	111.34	91.67	66.70	19.62	10.76
江 西 Jiangxi	4317	102.41	81.75	41.44	19.57	9.37
山 东 Shandong	3628	86.51	92.44	53.84	19.53	8.34
河 南 Henan	4458	102.43	83.46	26.83	20.46	6.08
湖 北 Hubei	2657	117.39	84.53	42.84	26.24	6.65
湖 南 Hunan	3895	107.89	73.73	28.36	19.90	8.86
广 东 Guangdong	2561	97.71	93.70	73.55	24.50	9.55
广 西 Guangxi	5964	100.90	89.28	58.78	22.36	9.12
海 南 Hainan	2634	95.51	87.67	74.86	24.57	3.78
重 庆 Chongqing	4695	88.70	91.47	44.73	18.18	12.48
四 川 Sichuan	4249	93.33	86.10	23.94	16.82	7.01
贵 州 Guizhou	3821	103.54	81.85	11.07	28.81	8.42
云 南 Yunnan	4535	96.03	94.47	9.58	19.86	10.49
西 藏 Tibet	4377	96.59	43.00	7.82	40.46	3.53
陕 西 Shaanxi	3087	92.34	83.44	17.81	27.42	7.09
甘 肃 Gansu	3367	73.45	93.28	8.61	18.75	6.94
青 海 Qinghai	4380	85.33	66.59	2.98	16.69	4.95
宁 夏 Ningxia	3224	87.99	94.29	14.32	26.56	7.05
新 疆 Xinjiang	3322	89.21	91.11	13.25	29.28	11.34

Level of Municipal Public Facilities of Built-up Area of Townships(2020)

污水处理率 (%) Wastewater Treatment Rate (%)	污水处理厂 集中处理率 Centralized Treatment Rate of Wastewater Treatment Plants	人均公园 绿地面积 （平方米） Public Recreational Green Space Per Capita (m²)	绿 化 覆盖率 (%) Green Coverage Rate (%)	绿地率 (%) Green Space Rate (%)	生活垃圾 处 理 率 (%) Domestic Garbage Treatment Rate (%)	无害化 处理率 Domestic Garbage Harmless Treatment Rate	地区名称 Name of Regions
21.67	**13.43**	**1.76**	**15.04**	**8.49**	**78.60**	**48.46**	**全　国**
23.10	21.84	2.82	30.86	21.07	98.91	97.55	北　京
53.37	21.23	2.21	6.74	2.68	22.11		天　津
4.58	2.55	0.97	13.34	8.01	71.53	45.31	河　北
12.20	6.76	1.45	22.05	9.21	59.69	16.65	山　西
1.49	0.78	1.37	11.30	7.07	29.61	18.94	内 蒙 古
1.68	1.54	0.32	15.31	7.21	44.99	22.81	辽　宁
3.07	2.57	1.67	11.81	8.62	97.57	91.69	吉　林
1.03	0.81	0.67	7.24	4.76	38.54	30.63	黑 龙 江
67.23	67.23	2.46	26.27	22.03	100.00	100.00	上　海
78.29	67.15	5.71	30.21	23.62	99.88	96.42	江　苏
47.67	18.16	2.18	15.29	7.73	90.58	62.44	浙　江
46.58	35.42	3.18	19.94	11.67	97.18	91.31	安　徽
81.13	55.89	7.18	25.82	17.25	99.35	98.63	福　建
26.40	16.00	1.31	12.25	7.86	86.52	47.01	江　西
49.81	23.54	2.68	21.79	12.22	99.99	96.77	山　东
16.07	11.22	2.07	17.75	7.03	79.46	42.92	河　南
44.63	21.43	2.08	9.57	4.97	92.06	77.07	湖　北
6.95	3.08	2.10	22.12	12.36	76.42	38.70	湖　南
22.61	15.66	4.83	22.99	12.50	93.68	65.86	广　东
8.11	6.52	2.84	15.25	10.73	96.55	48.76	广　西
3.05	3.05	0.04	19.99	13.68	100.00	6.18	海　南
78.75	61.92	0.52	12.09	7.93	92.07	59.19	重　庆
24.11	20.45	0.48	8.42	5.61	79.44	43.57	四　川
20.86	15.33	1.02	11.28	6.65	83.61	44.82	贵　州
27.60	11.05	0.74	8.18	5.11	79.55	38.74	云　南
0.32	0.05	0.05	8.67	5.67	80.12	14.44	西　藏
30.45	28.18	0.25	5.56	3.72	84.45	19.64	陕　西
21.92	17.23	0.45	12.08	7.58	56.62	44.86	甘　肃
0.05	0.05	0.02	7.85	4.82	61.35	16.10	青　海
36.69	26.09	1.06	11.94	8.08	91.98	41.66	宁　夏
8.47	3.53	1.17	18.03	13.62	71.44	39.85	新　疆

3-2-10 乡基本情况(2020年)

地区名称 Name of Regions	乡个数 (个) Number of Townships (unit)	建成区 面积 (公顷) Surface Area of Built-up Districts (hectare)	建成区 户籍人口 (万人) Registered Permanent Population (10,000 persons)	建成区 常住人口 (万人) Permanant Population (10,000 persons)
全　国　**National Total**	**8876**	**616993.10**	**2371.60**	**2294.02**
北　京　Beijing	15	507.50	2.17	2.06
天　津　Tianjin	3	1084.25	0.78	0.91
河　北　Hebei	733	61707.46	205.48	196.93
山　西　Shanxi	595	41428.49	120.67	124.74
内 蒙 古　Inner Mongolia	259	21723.60	46.89	43.67
辽　宁　Liaoning	189	11060.59	38.95	38.52
吉　林　Jilin	165	12877.05	35.17	36.25
黑 龙 江　Heilongjiang	332	26357.20	74.41	62.79
上　海　Shanghai	2	136.38	0.56	0.47
江　苏　Jiangsu	36	5737.96	32.01	31.39
浙　江　Zhejiang	238	13029.14	54.93	47.40
安　徽　Anhui	260	30652.39	118.94	116.81
福　建　Fujian	256	15661.90	85.45	77.84
江　西　Jiangxi	547	42710.34	191.78	184.40
山　东　Shandong	67	8635.03	33.41	31.32
河　南　Henan	576	88012.92	400.84	392.36
湖　北　Hubei	159	33944.23	83.39	90.20
湖　南　Hunan	375	35492.03	144.82	138.24
广　东　Guangdong	14	1024.00	3.79	2.62
广　西　Guangxi	307	12465.19	78.24	74.35
海　南　Hainan	21	884.41	2.04	2.33
重　庆　Chongqing	174	6062.04	30.82	28.46
四　川　Sichuan	1052	22498.49	97.70	95.60
贵　州　Guizhou	294	23444.99	95.32	89.57
云　南　Yunnan	540	32065.77	138.64	145.43
西　藏　Tibet	528	7171.08	32.20	31.39
陕　西　Shaanxi	21	1314.37	5.26	4.06
甘　肃　Gansu	336	12668.52	44.11	42.65
青　海　Qinghai	224	6501.05	29.83	28.48
宁　夏　Ningxia	87	5300.73	19.39	17.09
新　疆　Xinjiang	471	34834.00	123.64	115.71

Summary of Townships(2020)

设有村镇建设管理机构的个数（个）Number of Towns with Construction Management Institution (unit)	规划建设管理　Planning and Administer		有总体规划的乡个数（个）Number of Townships with Master Plans (unit)		本年规划编制投入（万元）Input in Planning This Year (10,000 RMB)	地区名称 Name of Regions	
	村镇建设管理人员（人）Number of Construction Management Personnel (person)	专职人员 Full-time Staff		本年编制 Compiled This Year			
6692	19509	12364	6491	367	70751.74	全	国
15	77	39	13		1580.00	北	京
3	18	12			16.00	天	津
603	1655	1147	463	39	2182.50	河	北
386	949	561	294	7	4637.70	山	西
206	483	303	197	6	224.00	内 蒙 古	
186	313	274	157	4	254.90	辽	宁
160	428	313	91	6	316.60	吉	林
318	552	330	238	17	111.68	黑 龙 江	
2	27	17	2		0.00	上	海
36	197	157	35	3	376.50	江	苏
191	524	330	205	15	2213.30	浙	江
221	747	473	225	16	5492.63	安	徽
241	507	342	246	12	1839.35	福	建
523	1502	936	526	37	5213.09	江	西
65	253	180	59	7	358.00	山	东
557	2904	1849	466	16	10868.47	河	南
152	655	417	133	4	2859.20	湖	北
316	1132	678	292	9	5033.10	湖	南
11	52	19	13		347.00	广	东
302	724	492	284	4	812.17	广	西
21	36	53	21		79.00	海	南
173	468	346	163	12	550.70	重	庆
343	701	397	344	17	2131.41	四	川
290	617	389	274	33	1843.40	贵	州
508	1793	1068	493	31	6307.40	云	南
86	362	201	293	43	5975.80	西	藏
16	30	18	12	1	1320.00	陕	西
219	537	294	270	6	3249.00	甘	肃
133	157	89	162	11	391.00	青	海
80	151	83	78	6	196.00	宁	夏
329	958	557	442	5	3971.84	新	疆

3-2-11 乡供水(2020年)

地区名称 Name of Regions		集中供水 的乡个数 （个） Number of Townships with Access to Piped Water (unit)	占全部乡 的比例 （%） Percentage of Total Rate (%)	公共供水综合 生产能力 （万立方米／日） Integrated Production Capacity of Public Water Supply Facilities (10,000 m³/day)	自备水源单位 综合生产能力 （万立方米／日） Integrated Production Capacity of Self-built Water Supply Facilities (10,000 m³/day)	年供水总量 （万立方米） Annual Supply of Water (10,000 m³)
全　　国	**National Total**	**7869**	**88.65**	**1182.43**	**413.28**	**135263.31**
北　京	Beijing	15	100.00	2.79	0.75	201.92
天　津	Tianjin	3	100.00	0.36	0.20	44.98
河　北	Hebei	646	88.13	58.47	32.02	10076.03
山　西	Shanxi	554	93.11	31.80	10.29	6089.01
内　蒙　古	Inner Mongolia	233	89.96	16.98	5.54	1997.49
辽　宁	Liaoning	108	57.14	8.20	2.83	1389.48
吉　林	Jilin	163	98.79	21.64	12.97	2261.27
黑　龙　江	Heilongjiang	313	94.28	27.40	50.81	2295.30
上　海	Shanghai	2	100.00	0.96	0.09	63.89
江　苏	Jiangsu	36	100.00	9.49	1.73	1803.71
浙　江	Zhejiang	224	94.12	23.25	9.18	3749.63
安　徽	Anhui	256	98.46	56.94	12.16	6279.02
福　建	Fujian	250	97.66	27.11	8.31	5454.70
江　西	Jiangxi	533	97.44	94.90	31.74	10152.72
山　东	Shandong	65	97.01	13.04	5.17	2138.53
河　南	Henan	553	96.01	190.88	102.18	22772.51
湖　北	Hubei	149	93.71	62.86	21.53	5905.01
湖　南	Hunan	347	92.53	71.49	12.50	11032.63
广　东	Guangdong	12	85.71	2.32	0.43	328.07
广　西	Guangxi	300	97.72	15.20	4.14	3420.68
海　南	Hainan	21	100.00	0.91	0.10	108.34
重　庆	Chongqing	172	98.85	7.92	1.40	1388.41
四　川	Sichuan	850	80.80	69.55	23.84	5024.85
贵　州	Guizhou	275	93.54	22.71	13.37	4815.47
云　南	Yunnan	535	99.07	66.91	30.55	10116.77
西　藏	Tibet	200	37.88	132.43	4.33	6678.42
陕　西	Shaanxi	20	95.24	0.97	0.21	153.97
甘　肃	Gansu	331	98.51	47.21	2.47	1820.94
青　海	Qinghai	152	67.86	5.75	0.55	862.15
宁　夏	Ningxia	87	100.00	7.43	1.45	841.78
新　疆	Xinjiang	464	98.51	84.57	10.45	5995.63

Water Supply of Townships(2020)

年生活 用水量 Annual Domestic Water Consumption	年生产 用水量 Annual Water Consumption for Production	供水管道 长　度 （公里） Length of Water Supply Pipelines (km)	本年新增 Added This Year	用水人口 （万人） Population with Access to Water (10,000 persons)	地区名称 Name of Regions
68115.94	53386.31	146879.52	10089.73	1923.44	全　　国
89.49	48.39	232.50	2.00	1.90	北　京
20.85	24.13	47.31	0.00	0.72	天　津
5288.44	3931.07	6179.22	431.30	165.75	河　北
2893.60	2635.28	9223.97	235.50	97.13	山　西
923.29	860.77	2331.16	124.42	29.92	内　蒙　古
724.20	538.14	1679.33	120.93	20.70	辽　宁
944.39	1022.80	1891.66	335.57	30.91	吉　林
1643.12	576.68	3566.17	85.72	55.51	黑　龙　江
19.50	33.08	39.00	0.00	0.47	上　海
1024.99	704.97	1448.27	59.65	31.18	江　苏
1727.49	1736.92	4556.48	197.39	43.42	浙　江
3810.47	1949.12	7522.49	805.64	102.08	安　徽
2899.61	2056.01	4083.84	286.43	71.35	福　建
5634.89	3587.68	8951.98	752.11	150.75	江　西
914.31	1138.57	1271.05	94.90	28.96	山　东
12242.66	8449.00	19308.72	1475.14	327.45	河　南
3266.93	2165.76	7731.01	491.46	76.24	湖　北
4013.98	6094.45	8292.24	399.22	101.93	湖　南
87.63	160.24	179.90	25.72	2.46	广　东
2444.67	748.67	3024.18	270.99	66.38	广　西
71.18	35.26	147.54	4.80	2.04	海　南
842.85	409.40	1320.02	50.97	26.03	重　庆
2803.84	1539.63	9946.36	816.41	82.31	四　川
2770.85	1222.52	6871.17	660.62	73.32	贵　州
4815.47	4615.90	14409.43	1095.55	137.38	云　南
475.88	4133.00	2237.57	186.98	13.50	西　藏
114.12	25.43	345.13	27.22	3.39	陕　西
1066.65	586.74	3124.62	242.74	39.78	甘　肃
590.59	266.29	1425.17	24.73	18.96	青　海
517.45	266.98	1059.81	48.92	16.11	宁　夏
3432.55	1823.43	14432.22	736.70	105.42	新　疆

3-2-12 乡燃气、供热、道路桥梁(2020年)

地区名称 Name of Regions	用气人口 (万人) Population with Access to Gas (10,000 persons)	集中供热 (万平方米) Area of Centrally Heated District (10,000 m²)	道路长度 (公里) Length of Roads (km)	本年新增 Added This Year	本年更 新改造 Renewal and Upgrading during The Reported Year
全 国 **National Total**	**707.74**	**3040.41**	**88978.59**	**7025.63**	**5707.50**
北 京 Beijing	0.81	21.09	74.48	0.03	5.73
天 津 Tianjin	0.59	6.35	36.51		
河 北 Hebei	89.68	567.19	6256.46	72.37	366.26
山 西 Shanxi	24.22	458.24	4297.25	198.25	252.49
内 蒙 古 Inner Mongolia	8.52	201.13	2095.75	29.30	74.57
辽 宁 Liaoning	5.65	149.71	1697.78	83.20	162.38
吉 林 Jilin	15.51	115.91	1403.64	17.16	137.75
黑 龙 江 Heilongjiang	5.18	139.43	2750.51	6.50	46.59
上 海 Shanghai	0.08		30.31	2.00	2.00
江 苏 Jiangsu	30.40		1282.56	50.89	78.93
浙 江 Zhejiang	23.60	28.80	2057.24	143.71	154.91
安 徽 Anhui	42.86	44.00	4799.04	567.26	232.33
福 建 Fujian	51.92		2823.47	243.32	152.43
江 西 Jiangxi	76.41	57.70	6363.41	661.77	499.20
山 东 Shandong	16.86	139.74	975.72	74.20	88.91
河 南 Henan	105.28	197.31	13679.39	1546.87	859.32
湖 北 Hubei	38.64	18.96	4011.54	241.52	367.74
湖 南 Hunan	39.21	1.00	4675.96	404.62	416.12
广 东 Guangdong	1.93		145.01	51.19	11.36
广 西 Guangxi	43.70		2817.92	169.15	132.11
海 南 Hainan	1.74		94.79	5.00	8.00
重 庆 Chongqing	12.73		938.07	40.98	142.99
四 川 Sichuan	22.88		3351.87	359.99	179.05
贵 州 Guizhou	9.91	53.35	4411.12	411.72	258.90
云 南 Yunnan	13.94	6.21	5616.02	550.23	400.44
西 藏 Tibet	2.46	37.50	2723.98	286.65	196.60
陕 西 Shaanxi	0.72	10.82	255.82	30.20	12.10
甘 肃 Gansu	3.67	114.88	1547.55	103.44	65.95
青 海 Qinghai	0.85	51.33	921.79	47.10	75.55
宁 夏 Ningxia	2.45	85.45	776.46	73.25	43.39
新 疆 Xinjiang	15.33	534.31	6067.17	553.76	283.40

Gas, Central Heating, Roads and Bridges of Townships(2020)

安装路灯的 道路长度 Roads with Lamps	道路面积 （万平方米） Surface Area of Roads (10,000 m²)	本年新增 Added This year	本年更 新改造 Renewal and Upgrading during The Reported Year	桥梁座数 （座） Number of Bridges (unit)	本年新增 Added This year	地区名称 Name of Regions
22640.30	**49123.89**	**4455.84**	**3058.85**	**17220**	**1448**	全　国
59.82	34.17	0.01	1.13	24		北　京
17.00	16.02			2		天　津
1815.58	3104.75	117.58	177.42	999	85	河　北
2015.32	2249.57	155.93	140.95	563	13	山　西
400.29	1077.72	28.21	28.89	201	13	内　蒙　古
364.47	913.03	58.61	84.99	421	47	辽　宁
402.41	744.35	22.05	59.23	219	3	吉　林
627.29	1492.29	16.56	35.37	232	14	黑　龙　江
2.56	13.14	0.80	0.80	10		上　海
541.07	730.66	30.78	42.89	219	17	江　苏
501.19	1279.70	65.13	59.61	1050	30	浙　江
1525.76	2698.73	320.26	99.30	1147	63	安　徽
922.01	1527.29	140.59	76.28	827	28	福　建
1395.77	3608.57	406.75	239.81	1468	115	江　西
377.56	611.76	42.07	46.71	363	38	山　东
3059.26	8027.88	1021.36	557.79	2560	240	河　南
1086.14	2366.56	192.99	188.16	756	37	湖　北
1199.63	2751.42	220.90	363.27	1278	289	湖　南
40.19	64.25	21.28	4.53	31	1	广　东
822.50	1662.30	120.20	70.77	477	43	广　西
60.80	57.24	0.98	1.50	27		海　南
310.92	517.43	28.76	48.52	272	15	重　庆
629.00	1608.43	153.76	84.43	841	68	四　川
919.41	2580.30	330.05	200.38	354	38	贵　州
1163.54	2888.63	287.42	152.93	693	63	云　南
420.27	1269.84	194.27	82.19	949	91	西　藏
21.87	111.26	8.25	3.65	15		陕　西
292.85	799.72	57.10	30.91	301	33	甘　肃
135.41	475.29	21.16	23.57	237	13	青　海
216.93	453.91	50.63	23.89	103	7	宁　夏
1293.48	3387.68	341.40	128.98	581	44	新　疆

3-2-13 乡排水和污水处理(2020年)
Drainage and Wastewater Treatment of Townships(2020)

地区名称 Name of Regions	对生活污水进行处理的乡 Township with Domestic Wastewater Treated		污水处理厂 Wastewater Treatment Plant		污水处理装置处理能力(万立方米/日) Treatment Capacity of Wastewater Treatment Facilities (10,000 m³/day)	排水管道长度(公里) Length of Drainage Pipelines (km)	本年新增 Added This Year	排水暗渠长度(公里) Length of Drains (km)	本年新增 Added This Year
	个数(个) Number (unit)	比例(%) Rate (%)	个数(个) Number of Wastwater Treatment Plants (unit)	处理能力(万立方米/日) Treatment Capacity (10,000 m³/day)					
全 国 National Total	3095	34.87	2170	104.80	98.80	23536.98	2435.74	20770.38	1577.15
北 京 Beijing	10	66.67	3	0.03	0.82	19.70	0.00	16.35	0.50
天 津 Tianjin	2	66.67	2	0.05	0.09	21.90	10.20	2.86	0.00
河 北 Hebei	97	13.23	17	3.55	5.32	1306.48	110.95	747.86	49.67
山 西 Shanxi	106	17.82	22	3.53	3.97	989.31	98.34	707.17	22.81
内 蒙 古 Inner Mongolia	28	10.81	6	0.32	0.05	400.20	60.60	307.61	34.37
辽 宁 Liaoning	29	15.34	10	0.59	0.56	220.74	22.50	190.74	7.60
吉 林 Jilin	30	18.18	8	0.54	1.18	282.95	28.06	250.83	25.98
黑 龙 江 Heilongjiang	19	5.72	7	0.09	0.09	446.00	9.43	495.88	5.30
上 海 Shanghai	2	100.00	2	0.11	0.11	15.00	3.00	6.00	3.00
江 苏 Jiangsu	35	97.22	41	2.58	2.25	633.69	30.11	276.34	10.83
浙 江 Zhejiang	218	91.60	31	3.65	3.36	844.67	80.04	933.26	27.10
安 徽 Anhui	223	85.77	191	7.86	7.74	1542.39	198.97	1678.79	173.48
福 建 Fujian	249	97.27	302	16.65	13.54	1064.78	123.46	620.83	46.00
江 西 Jiangxi	269	49.18	183	5.76	4.72	2206.06	251.61	1795.81	177.45
山 东 Shandong	49	73.13	35	4.18	4.33	423.59	36.48	296.67	11.80
河 南 Henan	213	36.98	140	12.28	14.92	2996.58	395.66	2354.59	297.90
湖 北 Hubei	144	90.57	95	8.24	8.32	1188.74	103.01	1067.92	58.06
湖 南 Hunan	168	44.80	62	3.54	1.94	1306.88	124.20	1837.13	125.75
广 东 Guangdong	12	85.71	7	0.67	0.57	48.76	5.39	49.01	1.90
广 西 Guangxi	60	19.54	50	1.51	1.43	643.64	44.98	492.95	47.10
海 南 Hainan	1	4.76	1		0.01	22.75	0.60	10.65	0.00
重 庆 Chongqing	163	93.68	169	5.23	2.99	457.20	59.23	299.29	17.01
四 川 Sichuan	252	23.95	439	8.76	7.83	820.29	103.16	755.75	66.25
贵 州 Guizhou	155	52.72	75	7.16	5.97	904.01	128.19	1070.34	82.21
云 南 Yunnan	229	42.41	154	3.61	3.11	1450.88	129.49	1914.29	152.48
西 藏 Tibet	21	3.98	2	0.13	0.01	90.60	16.44	162.48	26.07
陕 西 Shaanxi	10	47.62	5	0.08	0.08	46.09	1.11	47.04	3.37
甘 肃 Gansu	108	32.14	29	1.04	1.23	466.53	52.12	413.14	37.80
青 海 Qinghai	23	10.27	2	0.13	0.08	187.39	11.15	134.61	12.41
宁 夏 Ningxia	42	48.28	42	1.45	0.50	254.76	21.63	118.79	15.09
新 疆 Xinjiang	128	27.18	38	1.48	1.66	2234.42	175.63	1715.40	37.86

3-2-14 乡园林绿化及环境卫生(2020年)
Landscaping and Environmental Sanitation of Townships(2020)

地区名称 Name of Regions	园林绿化(公顷) Landscaping(hectare)				环境卫生 Environmental Sanitation		
	绿化覆盖面积 Green Coverage Area	绿地面积 Area of Parks and Green Space	本年新增 Added This Year	公园绿地面积 Public Green Space	生活垃圾中转站 (座) Number of Garbage Transfer Station (unit)	环卫专用车辆设备 (辆) Number of Special Vehicles for Environmental Sanitation (unit)	公共厕所 (座) Number of Latrines (unit)
全 国 National Total	92784.79	52364.08	4255.30	4044.93	8947	28560	38802
北 京 Beijing	156.61	106.91	18.01	5.80	24	72	139
天 津 Tianjin	73.12	29.07	4.00	2.00	7	14	16
河 北 Hebei	8233.27	4944.47	423.45	191.07	524	2477	2805
山 西 Shanxi	9133.16	3816.46	210.29	180.91	336	2338	2143
内 蒙 古 Inner Mongolia	2454.09	1535.06	119.61	59.65	163	873	1246
辽 宁 Liaoning	1693.79	798.00	112.89	12.36	109	611	530
吉 林 Jilin	1520.27	1109.42	119.67	60.64	163	509	326
黑 龙 江 Heilongjiang	1908.69	1254.57	139.67	41.82	113	947	557
上 海 Shanghai	35.83	30.05	2.00	1.16	4	15	10
江 苏 Jiangsu	1733.18	1355.17	56.43	179.38	45	362	244
浙 江 Zhejiang	1992.21	1007.15	91.92	103.11	316	937	2130
安 徽 Anhui	6111.07	3577.05	240.77	371.62	279	1434	1618
福 建 Fujian	4044.04	2701.68	154.38	558.55	254	925	1820
江 西 Jiangxi	5233.19	3357.26	214.70	241.96	1622	1522	2397
山 东 Shandong	1881.90	1055.16	40.94	83.96	70	261	315
河 南 Henan	15625.43	6184.84	950.90	813.25	913	3483	3744
湖 北 Hubei	3248.19	1685.84	180.88	187.74	224	954	1669
湖 南 Hunan	7850.12	4387.91	181.81	290.88	903	1067	2221
广 东 Guangdong	235.43	127.98	20.74	12.66	11	79	115
广 西 Guangxi	1900.44	1337.06	121.29	211.27	265	1008	624
海 南 Hainan	176.75	120.95	4.00	0.10	19	40	21
重 庆 Chongqing	733.00	480.85	18.79	14.70	141	446	479
四 川 Sichuan	1894.24	1263.05	52.45	46.21	756	1448	2372
贵 州 Guizhou	2644.83	1560.25	99.11	90.92	300	1115	1507
云 南 Yunnan	2624.26	1637.16	141.61	107.82	261	1338	3165
西 藏 Tibet	621.51	406.76	45.42	1.65	418	1000	2282
陕 西 Shaanxi	73.08	48.95	3.19	1.02	18	56	125
甘 肃 Gansu	1530.69	960.03	154.94	19.30	290	755	900
青 海 Qinghai	510.07	313.33	39.02	0.51	107	627	858
宁 夏 Ningxia	633.11	428.52	44.96	18.06	83	356	253
新 疆 Xinjiang	6279.22	4743.12	247.46	134.85	209	1491	2171

3-2-15 乡房屋(2020年)

地区名称 Name of Regions	住宅　Residential Building						公共建筑	
	年末实有建筑面积(万平方米) Total Floor Space of Buildings (year-end) (10,000 m²)	混合结构以上 Mixed Strucure and Above	本年竣工建筑面积(万平方米) Floor Space Completed This Year (10,000 m²)	混合结构以上 Mixed Strucure and Above	房地产开发 Real Estate Development	人均住宅建筑面积(平方米) Per Capita Floor Space (m²)	年末实有建筑面积(万平方米) Total Floor Space of Buildings (year-end) (10,000 m²)	混合结构以上 Mixed Strucure and Above
全　国 National Total	83897.72	59513.12	4374.39	3641.80	446.42	35.38	22644.64	18930.33
北　京 Beijing	135.90	135.86	1.22	1.22	0.00	62.65	15.06	15.05
天　津 Tianjin	37.22	21.62	0.00	0.00	0.00	47.64	4.15	3.73
河　北 Hebei	6344.16	5061.08	134.17	97.43	11.08	30.88	1452.66	1348.58
山　西 Shanxi	3780.58	2237.88	76.10	59.53	8.30	31.33	859.77	521.25
内 蒙 古 Inner Mongolia	1246.74	1036.24	10.73	10.42	0.00	26.59	704.72	685.28
辽　宁 Liaoning	1003.57	676.42	75.94	75.65	5.05	25.77	360.32	303.44
吉　林 Jilin	934.48	717.28	4.40	4.16	0.08	26.57	245.20	238.40
黑 龙 江 Heilongjiang	1880.79	1702.20	7.60	6.81	0.00	25.28	422.89	413.86
上　海 Shanghai	15.73	14.00	0.00	0.00	0.00	28.24	7.08	6.42
江　苏 Jiangsu	1307.63	1115.24	63.78	63.74	26.27	40.85	351.42	308.06
浙　江 Zhejiang	2749.34	1840.30	47.78	46.22	4.64	50.06	327.46	295.65
安　徽 Anhui	3610.60	2891.40	105.74	94.50	24.46	30.36	871.41	754.78
福　建 Fujian	3356.53	2518.98	127.69	106.71	32.82	39.28	1011.69	891.27
江　西 Jiangxi	6775.44	5717.66	255.55	230.69	24.75	35.33	1900.74	1409.27
山　东 Shandong	1141.57	815.64	345.61	291.18	33.64	34.17	399.71	377.24
河　南 Henan	13768.05	10271.38	1155.21	753.21	233.30	34.35	2506.03	2019.25
湖　北 Hubei	3036.08	2366.30	115.17	101.42	8.11	36.41	638.49	522.61
湖　南 Hunan	5858.90	2963.68	150.88	126.49	6.61	40.46	702.08	611.01
广　东 Guangdong	164.73	134.26	18.12	5.76	0.00	43.44	16.50	15.84
广　西 Guangxi	2501.95	1986.40	77.09	66.58	1.81	31.98	639.78	593.32
海　南 Hainan	55.29	55.09	0.09	0.09	0.00	27.08	20.58	20.56
重　庆 Chongqing	1061.57	876.05	19.04	18.48	0.60	34.45	203.68	185.96
四　川 Sichuan	3391.70	2524.55	94.89	76.52	7.20	34.72	738.82	665.46
贵　州 Guizhou	3936.27	2818.38	842.73	815.56	7.20	41.30	541.55	438.64
云　南 Yunnan	5070.43	3664.22	270.57	254.92	4.34	36.57	1266.81	1140.28
西　藏 Tibet	4266.22	587.51	51.30	44.68	0.48	132.49	444.06	269.77
陕　西 Shaanxi	140.30	112.14	17.92	10.88	0.00	26.69	36.46	34.10
甘　肃 Gansu	1263.55	896.52	31.77	29.33	0.08	28.64	1766.34	1306.87
青　海 Qinghai	905.26	699.31	15.93	14.51	3.45	30.35	211.29	200.25
宁　夏 Ningxia	518.28	420.59	22.90	22.90	0.16	26.73	211.06	195.41
新　疆 Xinjiang	3638.87	2634.93	234.49	212.21	2.00	29.43	3766.83	3138.73

Building Construction of Townships(2020)

Public Building			生产性建筑 Industrial Building					
本年竣工建筑面积 (万平方米) Floor Space Completed This Year (10,000 m²)	混合结构以上 Mixed Strucure and Above	房地产开发 Real Estate Development	年末实有建筑面积 (万平方米) Total Floor Space of Buildings (year-end) (10,000 m²)	混合结构以上 Mixed Strucure and Above	本年竣工建筑面积 (万平方米) Floor Space Completed This Year (10,000 m²)	混合结构以上 Mixed Strucure and Above	房地产开发 Real Estate Development	地区名称 Name of Regions
1000.47	840.86	131.86	16312.25	11952.48	845.16	677.45	126.16	全　国
0.02	0.02	0.00	11.15	11.15	0.00	0.00	0.00	北　京
0.00	0.00	0.00	7.64	6.65	0.00	0.00	0.00	天　津
52.12	47.60	33.36	2010.03	1675.67	43.15	39.91	14.17	河　北
15.23	14.41	0.00	789.41	492.38	37.14	20.63	0.00	山　西
5.64	4.56	0.32	728.50	524.79	2.45	1.89	0.00	内 蒙 古
0.68	0.68	0.12	182.70	153.08	1.96	1.96	0.00	辽　宁
18.51	18.48	1.03	267.12	199.26	4.30	3.72	0.00	吉　林
1.30	1.30	0.10	288.85	269.04	0.33	0.20	0.00	黑 龙 江
0.12	0.12		8.55	8.55	0.00	0.00	0.00	上　海
7.29	7.29	1.00	576.60	514.63	19.53	18.02	0.00	江　苏
24.24	23.41	0.00	385.62	328.27	12.26	11.11	0.00	浙　江
32.97	26.58	4.25	542.38	457.32	46.54	37.72	0.18	安　徽
21.12	18.96	0.68	517.49	387.58	16.34	14.16	0.00	福　建
89.77	78.78	3.76	1076.58	890.27	53.78	48.85	1.43	江　西
23.70	22.34	1.39	362.99	314.78	27.19	24.81	1.90	山　东
253.22	182.92	52.56	2382.00	1953.52	201.06	127.03	26.58	河　南
40.90	35.81	5.20	461.64	345.86	32.27	23.12	7.02	湖　北
59.37	55.31	3.04	1159.80	331.85	31.41	29.28	0.48	湖　南
2.20	2.20	0.00	11.60	10.49	0.25	0.25	0.00	广　东
19.76	17.67	0.20	326.79	205.85	18.84	9.30	0.26	广　西
0.85	0.85	0.00	1.59	1.59	0.00	0.00	0.00	海　南
4.35	3.59	0.00	75.60	60.95	1.98	1.37	0.00	重　庆
45.63	37.91	4.29	1374.97	1230.26	11.38	9.37	0.06	四　川
57.77	37.97	13.54	262.58	175.50	17.47	13.50	0.20	贵　州
81.94	72.64	1.32	792.76	496.70	46.34	38.03	2.28	云　南
21.61	18.86	0.00	590.66	52.50	7.78	4.66	0.00	西　藏
3.25	3.25	0.00	16.78	11.04	0.61	0.59	0.00	陕　西
12.58	11.72	0.01	349.15	274.07	83.18	81.16	0.00	甘　肃
5.12	4.83	0.00	36.72	33.03	3.76	2.14	0.00	青　海
8.66	7.48	0.20	184.13	143.64	4.68	2.30	0.10	宁　夏
90.56	83.30	5.50	529.85	392.22	119.18	112.38	71.50	新　疆

3-2-16 乡建设投入(2020年)

计量单位: 万元

| 地区名称
Name of Regions | 本年建设
投入合计
Total
Construction
Input of
This Year | 房屋 Building | | | | | 小 计
Input of
Municipal
Public
Facilities | 供 水
Water
Supply |
		小 计 Input of House	房地产开发 Real Estate Development	住 宅 Residential Building	公共建筑 Public Building	生产性 建 筑 Industrial Building		
全 国 **National Total**	**7799565**	**6091686**	**583408**	**3637473**	**1675251**	**778961**	**1707879**	**211354**
北 京 Beijing	6825	1877	10	1807	70		4948	272
天 津 Tianjin	1239						1239	
河 北 Hebei	354724	305566	8486	183405	71720	50440	49159	3409
山 西 Shanxi	159937	106986	6713	75310	18743	12933	52950	3753
内 蒙 古 Inner Mongolia	41132	22118	675	10610	9214	2294	19014	4433
辽 宁 Liaoning	58975	49922	20885	30919	1611	17392	9053	962
吉 林 Jilin	44306	30106	70	4452	21134	4521	14200	2455
黑 龙 江 Heilongjiang	24389	9930	70	7355	2425	150	14459	202
上 海 Shanghai	734	225			225		509	
江 苏 Jiangsu	205261	164188	58795	127878	10366	25944	41072	3007
浙 江 Zhejiang	211618	122184	9208	61332	41964	18887	89435	15020
安 徽 Anhui	398660	273361	56982	189726	54159	29476	125299	15608
福 建 Fujian	383559	280058	174559	231339	29481	19238	103502	9096
江 西 Jiangxi	524175	375099	23839	243364	91790	39945	149076	13508
山 东 Shandong	514710	490132	74272	429654	26634	33845	24578	2681
河 南 Henan	1091170	897964	68082	584913	191879	121172	193206	19357
湖 北 Hubei	253310	149709	7994	71085	54163	24460	103602	7669
湖 南 Hunan	309524	227393	9190	151979	53777	21637	82131	12358
广 东 Guangdong	16812	9494		7321	1767	406	7318	207
广 西 Guangxi	181100	118761	6440	81087	23266	14409	62339	24174
海 南 Hainan	3883	1129		127	1002		2755	69
重 庆 Chongqing	67723	39922	2291	29418	9423	1082	27800	1694
四 川 Sichuan	360920	276776	28534	180459	82266	14051	84143	4100
贵 州 Guizhou	301088	203850	5986	123839	68576	11436	97238	17496
云 南 Yunnan	652748	492007	7357	338174	94479	59354	160741	27282
西 藏 Tibet	704032	664807	4437	66182	589286	9340	39224	5929
陕 西 Shaanxi	19730	16787	129	12458	3627	702	2942	153
甘 肃 Gansu	287290	243417	250	32876	14395	196146	43873	4538
青 海 Qinghai	67079	42896		26367	11432	5097	24184	504
宁 夏 Ningxia	68496	49435	357	29896	13670	5869	19061	805
新 疆 Xinjiang	484416	425587	7796	304143	82706	38738	58829	10613

Construction Input of Townships(2020)

Measurement Unit: 10,000 RMB

燃 气 Gas Supply	集中供热 Central Heating	道路桥梁 Road and Bridge	排 水 Drainage	污水处理 Wastewater Treatment	园林绿化 Landscaping	环境卫生 Environmental Sanitation	垃圾处理 Garbage Treatment	其 他 Other	地区名称 Name of Regions
42854	34807	546896	375867	272921	146160	228868	128337	121073	全 国
10	914	618	556	540	368	1581	441	630	北 京
900	8		200	200		131	24		天 津
11369	4313	10062	5399	1716	4691	8459	3947	1456	河 北
3570	7146	20989	6358	2547	4105	5325	2140	1705	山 西
61	1196	4894	1265	652	2188	3394	2208	1583	内 蒙 古
6	664	2241	2871	2852	275	1727	1039	308	辽 宁
280	763	3558	2261	913	1091	2838	1489	954	吉 林
	488	4668	1152	399	881	5576	3295	1492	黑 龙 江
		200	90	90	85	134	120		上 海
683		12983	6685	4709	5832	5522	2506	6360	江 苏
612		20415	11483	8327	15940	11958	5154	14008	浙 江
2449		27499	45750	35037	7301	16290	9844	10403	安 徽
193		41792	24661	19837	8276	14941	10338	4542	福 建
808	8	47489	39062	28154	13214	21393	10638	13593	江 西
1560	1638	6176	2244	960	3073	4003	1442	3203	山 东
11178	5998	64925	27068	15752	28152	25008	11518	11520	河 南
2444		37569	33717	28836	6436	9501	6593	6265	湖 北
1586	31	16530	29811	25999	4680	10823	6640	6312	湖 南
		2803	2913	2861	449	862	239	84	广 东
77		17824	6920	2665	4938	5663	4242	2743	广 西
			2496	1578	35	155	16		海 南
1247	300	5311	12555	11672	1659	3470	1885	1564	重 庆
2165		32734	28907	23222	8695	5618	4118	1924	四 川
177	102	27519	21959	17240	5996	12981	5577	11007	贵 州
35	20	74732	24840	17458	7185	18226	12628	8421	云 南
1	2894	15855	4802	1669	1496	6626	5227	1622	西 藏
45	17	530	658	512	216	670	384	655	陕 西
116	2378	16264	12166	7319	1133	5318	3527	1962	甘 肃
7	545	12467	826	90	1795	6992	5407	1046	青 海
353	1894	3954	5681	3125	1243	1639	861	3492	宁 夏
922	3491	14296	10511	5990	4732	12043	4850	2220	新 疆

3-2-17 镇乡级特殊区域市政公用设施水平(2020年)

地区名称 Name of Regions	人口密度 (人/平方公里) Population Density (person/km²)	人均日生 活用水量 (升) Per Capita Daily Water Consumption (liter)	供 水 普及率 (%) Water Coverage Rate (%)	燃 气 普及率 (%) Gas Coverage Rate (%)	人均道 路面积 (平方米) Road Surface Area Per Capita (m²)	排水管道 暗渠密度 (公里/平方公里) Density of Drains (km/km²)
全 国 National Total	**3115**	**120.70**	**92.39**	**66.28**	**20.70**	**6.61**
北 京 Beijing	3742	43.27	100.00	36.44	7.09	4.85
河 北 Hebei	2873	89.60	88.54	62.94	25.99	5.42
山 西 Shanxi	3929	84.37	96.73		95.42	8.48
内 蒙 古 Inner Mongolia	1508	81.63	73.01	4.35	58.83	2.59
辽 宁 Liaoning	3149	77.50	73.08	49.24	21.15	3.92
吉 林 Jilin	5020	80.82	99.95	12.28	23.95	10.95
黑 龙 江 Heilongjiang	471	83.61	73.12	16.96	74.81	1.46
上 海 Shanghai	2806	85.24	88.53	62.76	13.09	1.68
江 苏 Jiangsu	4525	92.31	98.57	84.89	20.38	11.10
安 徽 Anhui	7303	99.57	78.17	52.88	15.95	18.12
福 建 Fujian	5807	97.45	63.38	75.04	20.08	10.76
江 西 Jiangxi	3483	99.75	84.58	60.35	15.78	5.76
山 东 Shandong	1768	80.69	96.10	64.96	22.15	6.12
河 南 Henan	2852	63.82	38.57	30.09	81.63	6.25
湖 北 Hubei	3072	124.76	93.13	72.72	26.32	9.51
湖 南 Hunan	5355	99.09	63.66	30.26	13.53	5.00
广 东 Guangdong	2257	128.14	71.60	49.43	33.84	7.35
广 西 Guangxi	2081	92.91	58.68	50.30	14.75	3.55
海 南 Hainan	5515	170.14	91.50	17.97	14.60	9.46
云 南 Yunnan	4588	106.06	99.88	8.25	27.12	8.23
甘 肃 Gansu	3970	167.31	100.00		27.10	9.24
宁 夏 Ningxia	4095	90.49	93.11	51.13	17.64	6.80
新 疆 Xinjiang	2817	98.69	96.24	34.52	53.03	32.14
新疆生产 建设兵团 Xinjiang Production and Construction Corps	3301	130.88	97.16	77.24	17.43	6.16

Level of Municipal Public Facilities of Built-up Area of Special District at Township Level(2020)

污水处理率 (%) Wastewater Treatment Rate (%)	污水处理厂集中处理率 Centralized Treatment Rate of Wastewater Treatment Plants	人均公园绿地面积(平方米) Public Recreational Green Space Per Capita (m²)	绿化覆盖率 (%) Green Coverage Rate (%)	绿地率 (%) Green Space Rate (%)	生活垃圾处理率 (%) Domestic Garbage Treatment Rate (%)	无害化处理率 Domestic Garbage Harmless Treatment Rate	地区名称 Name of Regions
63.14	50.40	3.86	20.23	14.51	84.35	47.48	全　国
			96.98	96.97	100.00	100.00	北　京
61.37	57.22	0.36	15.05	10.15	82.30	61.53	河　北
		5.54	10.15	3.63	39.82	34.60	山　西
		0.24	14.74	7.24	25.34	18.09	内 蒙 古
		0.69	11.58	7.80	65.82	49.68	辽　宁
		1.93	13.16	5.73	100.00	100.00	吉　林
		8.50	15.00	3.90			黑 龙 江
56.04	56.04	10.73	37.05	32.73	100.00	100.00	上　海
75.87	55.38	10.78	24.21	18.92	99.49	56.25	江　苏
9.09	9.09	2.60	26.29	13.50	100.00	100.00	安　徽
74.97	56.23	2.59	22.99	19.22	98.19	98.07	福　建
38.84	19.27	0.48	8.82	6.79	93.66	41.70	江　西
15.25	12.85	6.48	16.87	13.32	100.00	100.00	山　东
6.00	4.31	1.11	18.29	4.38	45.75	43.25	河　南
61.68	48.30	3.13	24.72	14.37	92.20	74.78	湖　北
3.51		5.58	24.09	18.12	56.62	20.90	湖　南
9.49	8.54	0.62	16.22	8.24	99.87	43.95	广　东
76.15	76.15		23.73	10.19	100.00	93.02	广　西
63.93	63.93	0.54	9.82	4.49	99.97	5.94	海　南
		0.74	9.14	6.76	56.88	23.65	云　南
93.33	93.33	11.45	33.33	29.55	100.00	100.00	甘　肃
21.91	12.51	0.90	17.27	13.22	81.55	63.66	宁　夏
81.98	0.84	0.59	19.38	10.25	65.87	31.13	新　疆
75.34	70.30	4.42	21.16	16.36	85.11	41.96	新疆生产建设兵团

3-2-18 镇乡级特殊区域基本情况(2020年)

地区名称 Name of Regions	镇乡级特殊 区域个数 (个) Number of Special District at Township Level (unit)	建成区 面 积 (公顷) Surface Area of Built-up Districts (hectare)	建 成 区 户籍人口 (万人) Registered Permanent Population (10,000 persons)	建 成 区 常住人口 (万人) Permanant Population (10,000 persons)
全 国 National Total	447	72181.38	177.20	224.87
北 京 Beijing	1	132.00	0.35	0.49
河 北 Hebei	29	2624.14	7.71	7.54
山 西 Shanxi	6	137.90	0.94	0.54
内 蒙 古 Inner Mongolia	30	2591.10	3.77	3.91
辽 宁 Liaoning	22	1314.49	4.23	4.14
吉 林 Jilin	5	112.30	0.62	0.56
黑 龙 江 Heilongjiang	38	3848.71	3.28	1.81
上 海 Shanghai	2	2856.54	2.30	8.02
江 苏 Jiangsu	9	1087.05	4.57	4.92
安 徽 Anhui	13	743.45	4.68	5.43
福 建 Fujian	8	756.00	2.56	4.39
江 西 Jiangxi	24	2046.28	8.95	7.13
山 东 Shandong	6	2801.00	5.46	4.95
河 南 Henan	3	480.00	1.05	1.37
湖 北 Hubei	25	3381.85	10.69	10.39
湖 南 Hunan	13	257.58	1.31	1.38
广 东 Guangdong	9	570.30	1.45	1.29
广 西 Guangxi	3	1148.63	0.80	2.39
海 南 Hainan	6	1758.63	2.39	9.70
云 南 Yunnan	14	532.33	2.24	2.44
甘 肃 Gansu	1	6.60	0.03	0.03
宁 夏 Ningxia	15	1165.90	4.75	4.77
新 疆 Xinjiang	27	1670.42	4.12	4.71
新疆生产建设兵团 Xinjiang Production and Construction Corps	138	40158.18	98.93	132.58

Summary of Special District at Township Level(2020)

设有村镇建设管理机构的个数(个) Number of Towns with Construction Management Institution (unit)	村镇建设管理人员(人) Number of Construction Management Personnel (person)	规划建设管理 Planning and Administer		有总体规划的镇乡级特殊区域个数(个) Number of Special District at Township Level with Master Plans (unit)	本年编制 Compiled This Year	本年规划编制投入(万元) Input in Planning This Year (10,000 RMB)	地区名称 Name of Regions
		专职人员 Full-time Staff					
350	2475	1276		301	12	5334.08	全 国
1	6			1	0		北 京
22	49	44		15	1	99.00	河 北
2	2	1		0	0	0.00	山 西
27	93	79		22	0	10.00	内 蒙 古
21	35	30		15	0	1.60	辽 宁
4	5	5		2	0	0.00	吉 林
17	24	13		5	0	0.00	黑 龙 江
2	15	15		2	0		上 海
5	22	10		8	0		江 苏
10	188	114		9	0	0.00	安 徽
6	11	7		8	0	24.00	福 建
22	113	47		22	1	89.50	江 西
6	21	19		6	1	20.00	山 东
3	10	7		2	0	0.00	河 南
24	150	70		23	1	287.00	湖 北
9	15	12		4	0		湖 南
8	10	7		6	0	45.00	广 东
3	125	41		1	0		广 西
4	7	3		2	0		海 南
12	24	13		6	0	60.13	云 南
1	2	2		1	0		甘 肃
14	79	62		12	0	23.00	宁 夏
19	29	16		18	0	12.00	新 疆
108	1440	659		111	8	4662.85	新疆生产建设兵团

3-2-19 镇乡级特殊区域供水(2020年)

地区名称 Name of Regions	集中供水的 镇乡级特殊 区域个数 (个) Number of Special District at Township Level with Access to Piped Water (unit)	占全部 镇乡级特殊 区域的比例 (%) Percentage of Total Rate (%)	公共供水综合 生产能力 (万立方米 /日) Integrated Production Capacity of Public Water Supply Facilities (10,000 m³/day)	自备水源单位 综合生产能力 (万立方米/日) Integrated Production Capacity of Self-built Water Supply Facilities (10,000 m³/day)	年供水总量 (万立方米) Annual Supply of Water (10,000 m³)
全 国 National Total	386	86.35	103.64	21.16	16866.16
北 京 Beijing	1	100.00	0.45	0.45	99.60
河 北 Hebei	27	93.10	3.91	2.46	720.78
山 西 Shanxi	5	83.33	0.13	0.01	29.20
内 蒙 古 Inner Mongolia	27	90.00	1.78	0.37	143.99
辽 宁 Liaoning	16	72.73	1.64	0.04	153.88
吉 林 Jilin	5	100.00	0.48	0.09	20.53
黑 龙 江 Heilongjiang	16	42.11	0.77	0.07	53.58
上 海 Shanghai	2	100.00	1.45		390.93
江 苏 Jiangsu	9	100.00	2.27	0.29	349.48
安 徽 Anhui	13	100.00	1.59	0.16	217.86
福 建 Fujian	7	87.50	1.02	0.07	200.86
江 西 Jiangxi	20	83.33	1.51	0.49	418.14
山 东 Shandong	6	100.00	5.18	0.20	1202.95
河 南 Henan	2	66.67	0.15	0.14	46.00
湖 北 Hubei	25	100.00	8.09	0.94	694.19
湖 南 Hunan	12	92.31	0.47	0.04	47.52
广 东 Guangdong	6	66.67	0.49	1.12	119.03
广 西 Guangxi	3	100.00	2.33	0.50	207.21
海 南 Hainan	6	100.00	1.23	1.29	813.62
云 南 Yunnan	14	100.00	1.44	0.96	130.35
甘 肃 Gansu	1	100.00	0.01		1.60
宁 夏 Ningxia	15	100.00	4.15	0.68	525.59
新 疆 Xinjiang	24	88.89	9.00	0.30	1472.58
新疆生产 建设兵团 Xinjiang Production and Construction Corps	124	89.86	54.10	10.46	8806.69

Water Supply of Special District at Township Level(2020)

年生活 用水量 Annual Domestic Water Consumption	年生产 用水量 Annual Water Consumption for Production	供水管道 长 度 （公里） Length of Water Supply Pipelines (km)	本年新增 Added This Year	用水人口 （万人） Population with Access to Water (10,000 persons)	地区名称 Name of Regions
9152.42	7715.41	15502.63	269.58	207.75	全　国
7.80	64.00	30.80		0.49	北　京
218.29	451.82	639.98	11.00	6.67	河　北
16.14	10.00	74.50		0.52	山　西
85.00	46.78	262.60	13.00	2.85	内 蒙 古
85.57	54.17	142.90	0.00	3.03	辽　宁
16.62	3.20	49.00	0.00	0.56	吉　林
40.42	13.01	266.36	0.20	1.32	黑 龙 江
220.76	170.17	120.90	2.05	7.10	上　海
163.36	174.19	228.13	0.00	4.85	江　苏
154.24	56.99	249.12	7.50	4.24	安　徽
98.96	125.10	144.00	2.40	2.78	福　建
219.47	148.97	199.64	16.71	6.03	江　西
140.16	1047.49	148.00	0.00	4.76	山　东
12.30	26.00	95.00	0.00	0.53	河　南
440.61	203.53	668.32	10.50	9.68	湖　北
31.76	15.41	67.98	1.50	0.88	湖　南
43.10	74.93	163.90	5.20	0.92	广　东
47.56	159.65	40.20	3.00	1.40	广　西
551.10	225.50	259.40	0.00	8.87	海　南
94.44	25.84	2311.67	3.55	2.44	云　南
1.60		2.21		0.03	甘　肃
146.80	374.14	269.35	5.02	4.44	宁　夏
163.11	1293.53	816.27	15.51	4.53	新　疆
6153.25	2950.99	8252.40	172.44	128.81	新疆生产 建设兵团

3-2-20 镇乡级特殊区域燃气、供热、道路桥梁(2020年)

地区名称 Name of Regions	用气人口 (万人) Population with Access to Gas (10,000 persons)	集中供热 (万平方米) Area of Centrally Heated District (10,000 m²)	道路长度 (公里) Length of Roads (km)	本年新增 Added This Year	本年更新 改 造 Renewal and Upgrading during The Reported Year
全 国 National Total	**149.04**	**5314.97**	**6847.67**	**322.06**	**226.42**
北 京 Beijing	0.18	21.00	5.00		
河 北 Hebei	4.74	171.37	357.02	3.75	5.60
山 西 Shanxi			95.25		
内 蒙 古 Inner Mongolia	0.17	10.88	482.53	6.29	
辽 宁 Liaoning	2.04	53.50	174.77		11.07
吉 林 Jilin	0.07	1.35	31.80		
黑 龙 江 Heilongjiang	0.31	24.30	230.40		0.20
上 海 Shanghai	5.03		79.28	9.70	1.70
江 苏 Jiangsu	4.18		165.26	4.00	3.00
安 徽 Anhui	2.87		193.74	10.29	4.00
福 建 Fujian	3.29		115.78	6.35	
江 西 Jiangxi	4.30		213.20	16.87	30.30
山 东 Shandong	3.22	46.00	116.40		
河 南 Henan	0.41		275.70	54.40	9.42
湖 北 Hubei	7.56		423.13	31.26	16.32
湖 南 Hunan	0.42		40.43	4.20	7.96
广 东 Guangdong	0.64		95.12	7.56	
广 西 Guangxi	1.20		70.54		0.07
海 南 Hainan	1.74		175.25	7.86	
云 南 Yunnan	0.20		123.09	3.32	2.10
甘 肃 Gansu			2.00	1.00	
宁 夏 Ningxia	2.44	112.48	152.43	8.57	4.81
新 疆 Xinjiang	1.62	56.66	432.61	47.05	38.00
新疆生产 建设兵团 Xinjiang Production and Construction Corps	102.41	4817.43	2796.94	99.59	91.87

Gas, Central Heating, Road and Bridge of Special District at Township Level(2020)

安装路灯的 道路长度 Roads with Lamps	道路面积 (万平方米) Surface Area of Roads $(10,000 \text{ m}^2)$	本年新增 Added This year	本年更新 改　造 Renewal and Upgrading during The Reported Year	桥梁座数 (座) Number of Bridges (unit)	本年新增 Added This year	地区名称 Name of Regions
1737.49	4655.80	266.33	212.44	1309	104	全　　国
5.00	3.50					北　京
92.23	195.89	4.78	4.61	53	2	河　北
9.00	51.70			6		山　西
35.41	229.89	2.36		33		内 蒙 古
57.04	87.56		9.12	31		辽　宁
12.50	13.50					吉　林
21.63	135.51		0.30	8		黑 龙 江
57.70	104.94	23.28	1.78	42	12	上　海
34.88	100.23	2.00	2.10	32		江　苏
55.65	86.58	4.41		34	1	安　徽
30.25	88.16	8.14		19		福　建
47.30	112.43	7.93	16.50	28	4	江　西
69.70	109.69			25	1	山　东
11.30	111.75	22.30	15.00	4	1	河　南
119.94	273.49	14.34	63.22	180	1	湖　北
18.30	18.67	1.91	1.67	10		湖　南
16.07	43.55	2.08		16	1	广　东
3.30	35.26		0.01	5		广　西
18.90	141.58	3.62		21		海　南
27.50	66.24	14.11	12.04	19	1	云　南
2.00	0.71	0.19		1		甘　肃
42.96	84.21	3.38	1.91	20	1	宁　夏
97.53	249.53	27.49	21.60	121	77	新　疆
851.40	2311.23	124.01	62.58	601	2	新疆生产 建设兵团

3-2-21 镇乡级特殊区域排水和污水处理(2020年)
Drainage and Wastewater Treatment of Special District at Township Level(2020)

地区名称 Name of Regions	对生活污水进行处理的镇乡级特殊区域 Special District at Township Level with		污水处理厂 Wastewater Treatment Plant		污水处理装置处理能力（万立方米/日） Treatment Capacity of Wastewater Treatment Facilities (10,000 m³/day)	排水管道长度（公里） Length of Drainage Pipelines (km)		排水暗渠长度（公里） Length of Drains (km)	
	个数（个） Number (unit)	比例（%） Rate (%)	个数（个） Number of Wastwater Treatment Plants (unit)	处理能力（万立方米/日） Treatment Capacity (10,000 m³/day)			本年新增 Added This Year		本年新增 Added This Year
全　国 National Total	234	52.35	170	32.58	43.08	3571.20	148.53	1197.19	47.72
北　京 Beijing	0		0			6.40			
河　北 Hebei	10	34.48	7	6.91	2.22	111.82	3.22	30.49	1.28
山　西 Shanxi	2	33.33	2			6.20	1.00	5.50	0.50
内 蒙 古 Inner Mongolia	2	6.67	0	0.01		48.73		18.40	
辽　宁 Liaoning	1	4.55	1	0.07	0.07	27.59		23.90	0.80
吉　林 Jilin	1	20.00	0			4.50		7.80	2.50
黑 龙 江 Heilongjiang	0		0			37.92		18.43	
上　海 Shanghai	2	100.00	0			47.87	2.40		
江　苏 Jiangsu	7	77.78	3	0.55	0.71	91.60	1.30	29.06	1.10
安　徽 Anhui	5	38.46	0		0.06	53.76	1.32	80.98	4.18
福　建 Fujian	7	87.50	2	1.15	3.11	71.85	11.30	9.53	
江　西 Jiangxi	13	54.17	10	0.16	0.82	56.15	6.85	61.78	10.00
山　东 Shandong	5	83.33	3	0.35	0.26	116.50	1.00	55.00	1.00
河　南 Henan	2	66.67	1	0.01	0.01	21.00	5.00	9.02	0.01
湖　北 Hubei	23	92.00	17	1.09	0.88	174.70	8.40	146.99	9.18
湖　南 Hunan	3	23.08	0			10.77	1.10	2.10	0.09
广　东 Guangdong	5	55.56	1	0.03	0.03	33.30	0.60	8.60	0.10
广　西 Guangxi	1	33.33	1	1.00	1.00	24.68	7.00	16.05	
海　南 Hainan	1	16.67	0	2.45		112.14	2.51	54.29	1.10
云　南 Yunnan	5	35.71	0	0.00	0.00	31.39	2.00	12.42	1.40
甘　肃 Gansu	1	100.00	1	0.00	0.00	0.60		0.01	
宁　夏 Ningxia	11	73.33	8	0.27	0.16	72.47	3.24	6.76	
新　疆 Xinjiang	7	25.93	1	0.24	0.20	52.79	5.12	484.02	1.50
新疆生产建设兵团 Xinjiang Production and Construction Corps	120	86.96	112	18.33	33.55	2356.47	85.17	116.06	12.98

3-2-22 镇乡级特殊区域园林绿化及环境卫生(2020年)
Landscaping and Environmental Sanitation of Special District at Township Level(2020)

地区名称 Name of Regions	园林绿化(公顷) Landscaping((hectare)				环境卫生 Environmental Sanitation		
	绿化覆盖面积 Green Coverage Area	绿地面积 Area of Parks and Green Space	本年新增 Added This Year	公园绿地面积 Public Green Space	生活垃圾中转站(座) Number of Garbage Transfer Station (unit)	环卫专用车辆设备(辆) Number of Special Vehicles for Environmental Sanitation (unit)	公共厕所(座) Number of Latrines (unit)
全 国 National Total	14604.98	10473.94	712.95	868.58	325	1592	2011
北 京 Beijing	128.02	128.00	13.20		5	10	5
河 北 Hebei	395.02	266.26	20.82	2.75	17	79	114
山 西 Shanxi	14.00	5.00	0.50	3.00	9	42	10
内 蒙 古 Inner Mongolia	381.97	187.62	5.42	0.94	39	89	157
辽 宁 Liaoning	152.26	102.55	1.00	2.86	6	61	54
吉 林 Jilin	14.78	6.43	2.00	1.09	0	10	8
黑 龙 江 Heilongjiang	577.20	149.92	2.15	15.40	0	42	86
上 海 Shanghai	1058.28	934.92	280.93	86.00	1	28	11
江 苏 Jiangsu	263.20	205.72	10.87	53.03	9	41	43
安 徽 Anhui	195.43	100.40	6.01	14.10	26	38	102
福 建 Fujian	173.82	145.31	2.30	11.38	8	24	20
江 西 Jiangxi	180.53	138.97	5.15	3.40	20	31	75
山 东 Shandong	472.63	373.23	9.00	32.10	11	32	32
河 南 Henan	87.80	21.03	7.01	1.52	5	16	10
湖 北 Hubei	835.96	486.10	11.98	32.56	27	105	220
湖 南 Hunan	62.06	46.67	0.60	7.70	4	12	21
广 东 Guangdong	92.50	47.00	0.00	0.80	6	15	85
广 西 Guangxi	272.53	117.08	0.01	0.00	1	6	11
海 南 Hainan	172.67	78.94	0.15	5.25	3	13	13
云 南 Yunnan	48.67	35.96	1.10	1.80	1	12	69
甘 肃 Gansu	2.20	1.95	0.00	0.30		1	5
宁 夏 Ningxia	201.40	154.12	3.51	4.30	11	53	61
新 疆 Xinjiang	323.73	171.29	8.90	2.76	14	63	113
新疆生产建设兵团 Xinjiang Production and Construction Corps	8498.32	6569.47	320.34	585.54	102	769	686

3-2-23　镇乡级特殊区域房屋(2020年)

地区名称 Name of Regions	住宅　Residential Building						公共建筑	
	年末实有建筑面积(万平方米) Total Floor Space of Buildings (year-end) (10,000 m²)	混合结构以上 Mixed Strucure and Above	本年竣工建筑面积(万平方米) Floor Space Completed This Year (10,000 m²)	混合结构以上 Mixed Strucure and Above	房地产开发 Real Estate Development	人均住宅建筑面积(平方米) Per Capita Floor Space (m²)	年末实有建筑面积(万平方米) Total Floor Space of Buildings (year-end) (10,000 m²)	混合结构以上 Mixed Strucure and Above
全　国 National Total	8474.36	6714.65	578.16	200.86	418.75	47.83	2023.36	1849.64
北　京 Beijing	24.00	24.00				68.32	1.10	1.10
河　北 Hebei	324.40	279.58	3.59	0.45	0.00	42.08	97.10	91.72
山　西 Shanxi	10.19	3.39	0.04	0.01	0.03	10.81	1.89	
内 蒙 古 Inner Mongolia	105.94	74.75	0.18	0.18	0.00	28.09	23.84	22.45
辽　宁 Liaoning	105.44	86.13	0.08	0.08	0.00	24.95	30.61	29.97
吉　林 Jilin	26.76	17.94	0.00	0.00	0.00	42.84	1.94	1.91
黑 龙 江 Heilongjiang	96.64	94.99				29.45	15.46	14.66
上　海 Shanghai	360.63	360.63	116.08	116.08	113.08	156.91	117.59	117.59
江　苏 Jiangsu	168.36	131.21	1.80	1.80		36.86	54.68	41.80
安　徽 Anhui	224.97	195.89	3.00	3.00	0.00	48.07	28.27	28.23
福　建 Fujian	159.60	135.74	29.00	22.00	26.80	62.38	20.96	19.84
江　西 Jiangxi	217.64	162.57	11.46	10.36	0.00	24.31	15.63	14.28
山　东 Shandong	220.37	163.57	7.81	7.81	5.86	40.34	100.77	96.18
河　南 Henan	30.16	27.66	1.90	1.80	0.00	28.70	1.85	1.85
湖　北 Hubei	414.81	384.40	8.23	8.22	0.00	38.81	125.83	113.28
湖　南 Hunan	34.24	30.34	0.47	0.33	0.00	26.23	3.52	2.52
广　东 Guangdong	65.10	53.04	0.02	0.02	0.00	44.80	9.92	9.03
广　西 Guangxi	27.77	25.59	0.01	0.01	0.00	34.71	5.17	5.02
海　南 Hainan	71.28	71.28	10.34	0.64	9.70	29.82	16.55	15.97
云　南 Yunnan	76.32	63.16	0.72	0.72	0.00	34.01	79.21	71.38
甘　肃 Gansu	0.80	0.80	0.00	0.00	0.00	27.12	1.00	1.00
宁　夏 Ningxia	195.38	175.99	0.62	0.62	0.00	41.14	36.69	31.34
新　疆 Xinjiang	132.16	95.96	3.66	3.52	1.10	32.05	52.75	43.08
新疆生产建设兵团 Xinjiang Production and Construction Corps	5381.38	4056.04	379.14	23.21	262.18	54.39	1181.04	1075.41

Building Construction of Special District at Township Level(2020)

Public Building			生产性建筑 Industrial Building					地区名称
本年竣工建筑面积（万平方米）Floor Space Completed This Year (10,000 m²)	混合结构以上 Mixed Strucure and Above	房地产开发 Real Estate Development	年末实有建筑面积（万平方米）Total Floor Space of Buildings (year-end) (10,000 m²)	混合结构以上 Mixed Strucure and Above	本年竣工建筑面积（万平方米）Floor Space Completed This Year (10,000 m²)	混合结构以上 Mixed Strucure and Above	房地产开发 Real Estate Development	Name of Regions
122.62	105.60	59.94	2460.26	1808.10	155.28	145.65	39.66	全 国
			40.00	40.00				北 京
4.58	4.58	0.00	472.66	209.02	35.90	35.90	0.00	河 北
0.04	0.01	0.03						山 西
0.34	0.19	0.00	20.96	19.49	0.38	0.26	0.00	内 蒙 古
0.00	0.00	0.00	26.17	23.58	0.00	0.00	0.00	辽 宁
0.00	0.00	0.00	0.83	0.83	0.00	0.00	0.00	吉 林
0.38	0.00		40.25	22.25				黑 龙 江
62.16	62.16	54.16	233.50	233.50	71.00	71.00	34.00	上 海
0.02	0.02		106.86	95.04	0.69	0.69		江 苏
0.20	0.20	0.00	22.18	22.15	0.32	0.32	0.00	安 徽
0.80	0.80	0.00	216.02	215.53	9.20	9.20	0.00	福 建
2.12	1.62	0.00	28.36	22.82	2.65	2.45	0.00	江 西
2.17	1.72	0.50	399.49	332.26	23.64	15.21	5.00	山 东
0.20	0.20	0.00	0.58	0.50	0.00	0.00	0.00	河 南
3.02	3.00	0.00	99.51	71.67	1.85	1.25	0.00	湖 北
0.10	0.10	0.00	3.46	3.10	0.08	0.08	0.00	湖 南
0.00	0.00	0.00	0.00	0.00	0.00	0.00	0.00	广 东
0.00	0.00	0.00	26.35	17.63	0.00	0.00	0.00	广 西
8.17	8.17	2.73	7.38	4.76	1.72	1.72	0.00	海 南
2.57	2.51	0.00	8.49	7.43	1.09	0.85	0.00	云 南
0.00	0.00	0.00	0.15	0.15	0.00	0.00	0.00	甘 肃
0.23	0.23	0.00	37.19	36.96	0.00	0.00	0.00	宁 夏
0.32	0.32	0.00	15.64	8.73	0.40	0.36	0.00	新 疆
35.21	19.78	2.52	654.25	420.72	6.36	6.36	0.66	新疆生产建设兵团

3-2-24 镇乡级特殊区域建设投入(2020年)

计量单位: 万元

| 地区名称
Name of Regions | 合 计
Total
Construction
Input of
This Year | 房屋 Building | | | | | 小 计
Input of
Municipal
Public
Facilities | 供 水
Water
Supply |
		本年建设 投入小计 Input of House	房地产开发 Real Estate Development	住 宅 Residential Building	公共建筑 Public Building	生产性 建 筑 Industrial Building		
全 国 **National Total**	**2133018**	**1732419**	**1121444**	**920698**	**464334**	**347387**	**400599**	**26069**
北 京 Beijing	225						225	110
河 北 Hebei	94143	47164	7850	9079	9725	28360	46978	1737
山 西 Shanxi	100	100			100			
内 蒙 古 Inner Mongolia	1502	1220		374	532	314	282	43
辽 宁 Liaoning	1002	82		82			920	130
吉 林 Jilin	6						6	
黑 龙 江 Heilongjiang	5741	500			500		5241	17
上 海 Shanghai	997014	921256	708256	459616	248640	213000	75758	168
江 苏 Jiangsu	6180	3820		2900	20	900	2360	116
安 徽 Anhui	6533	4379		3325	605	449	2154	432
福 建 Fujian	87079	77250	57500	58350	2500	16400	9829	112
江 西 Jiangxi	31294	22217		14997	4900	2320	9077	539
山 东 Shandong	75422	58279	14982	15847	3539	38893	17143	342
河 南 Henan	3795	2810	1700	2180	630		985	50
湖 北 Hubei	32574	16999	3000	10112	4713	2174	15575	535
湖 南 Hunan	1735	550		324	148	77	1185	81
广 东 Guangdong	653	198		143	55		455	150
广 西 Guangxi	5253	34		34			5219	32
海 南 Hainan	110776	109274	25057	21373	83601	4300	1502	
云 南 Yunnan	13308	11439		1723	4420	5296	1869	26
甘 肃 Gansu	434	348	348		348		86	
宁 夏 Ningxia	4891	1421		736	684		3470	152
新 疆 Xinjiang	9975	6183	2860	5223	696	265	3792	643
新疆生产建设兵团 Xinjiang Production and Construction Corps	643384	446897	299892	314280	97978	34640	196486	20655

Construction Input of Special District at Township Level(2020)

Measurement Unit: 10,000 RMB

燃 气 Gas Supply	集中供热 Central Heating	道路桥梁 Road and Bridge	排 水 Drainage	污水处理 Wastewater Treatment	园林绿化 Landscaping	环境卫生 Environmental Sanitation	垃圾处理 Garbage Treatment	其 他 Other	地区名称 Name of Regions
7750	55105	137494	61273	43947	58155	30417	16212	24336	全 国
						115	29		北 京
868	5505	30483	4779	4665	2518	1052	758	36	河 北
									山 西
	28	10	12		46	142	75		内 蒙 古
	51	235	60		42	319	250	83	辽 宁
					2	4			吉 林
		5058			18	143	15	5	黑 龙 江
3290		41180	960		30060	100	60		上 海
28		456	458	300	185	430	201	687	江 苏
30		503	170		788	206	94	26	安 徽
20		7055	1558	1320	312	679	240	93	福 建
		4949	1484	1310	977	426	271	704	江 西
126	106	11510	742	298	1538	2068	1005	711	山 东
40		580	100	40	80	105	35	30	河 南
515		2937	7063	6013	1080	2239	1366	1207	湖 北
3		402	310	24	111	196	102	84	湖 南
		101				174	117	30	广 东
		5	5003	5003	16	164	114		广 西
		181	1168	1168	15	130	122	8	海 南
		1135	237	12	148	278	176	45	云 南
					39	25	25	23	甘 肃
	110	262	2085	2077	516	199	102	146	宁 夏
	205	735	605	5	546	1038	177	21	新 疆
2830	49099	29719	34480	21713	19119	20186	10878	20398	新疆生产建设兵团

3-2-25 村庄人口及面积(2020年)

地区名称 Name of Regions	村庄建设用地面积 (公顷) Area of Villages Construction Land (hectare)	行政村个数(个) Number of Administrative Villages(unit)		
		合 计 Total	500人以下 Under 500 (persons)	500-1000人 500-1000 (persons)
全　国　National Total	12731429.17	492995	68983	115430
北　京　Beijing	88177.61	3471	1057	1102
天　津　Tianjin	58983.87	2925	762	1104
河　北　Hebei	863843.24	44432	10012	14749
山　西　Shanxi	377797.62	21004	5858	7309
内 蒙 古　Inner Mongolia	260625.10	10994	2452	3676
辽　宁　Liaoning	472831.53	10730	389	1394
吉　林　Jilin	340545.72	9100	1287	2119
黑 龙 江　Heilongjiang	443088.74	8918	1087	1735
上　海　Shanghai	60221.95	1519	114	189
江　苏　Jiangsu	657329.06	13632	224	751
浙　江　Zhejiang	326545.17	16377	1341	4109
安　徽　Anhui	587131.51	14794	657	1475
福　建　Fujian	261152.68	13327	1369	3350
江　西　Jiangxi	454085.29	16905	1138	3131
山　东　Shandong	1045519.46	63856	19366	22345
河　南　Henan	997450.24	42069	2856	8946
湖　北　Hubei	507628.46	22110	1765	5755
湖　南　Hunan	665386.92	22159	774	3477
广　东　Guangdong	674636.04	18080	833	1999
广　西　Guangxi	504953.15	14174	269	1126
海　南　Hainan	107547.94	2746	167	496
重　庆　Chongqing	203895.72	8256	287	844
四　川　Sichuan	756175.62	29347	3059	4202
贵　州　Guizhou	350586.65	14294	948	2681
云　南　Yunnan	502578.59	13646	533	1272
西　藏　Tibet	93160.52	5142	3368	1368
陕　西　Shaanxi	363669.81	16218	1157	4188
甘　肃　Gansu	328272.01	15918	2219	5612
青　海　Qinghai	54986.05	4133	1377	1632
宁　夏　Ningxia	67128.14	2253	129	364
新　疆　Xinjiang	230330.25	8809	1102	2407
新疆生产 建设兵团　Xinjiang Production and Construction Corps	25164.51	1657	1027	523

Population and Area of Villages(2020)

1000人以上 Above 1000 (persons)	自然村个数 （个） Number of Natural Villages (unit)	村庄户籍 人　口 （万人） Registered Permanent Population (10,000 persons)	村庄常住 人　口 （万人） Permanart Population (10,000 persons)	地区名称 Name of Regions
308839	2362908	77671.33	67529.59	全　　国
1312	4552	327.94	470.85	北　京
1059	2947	237.40	238.69	天　津
19677	66013	4680.12	4243.35	河　北
7837	43305	1944.89	1677.07	山　西
4866	50096	1338.46	1044.39	内 蒙 古
8982	47489	1712.76	1576.10	辽　宁
5694	39038	1316.37	1133.93	吉　林
6083	34491	1666.93	1290.93	黑 龙 江
1216	18911	303.69	444.23	上　海
12657	121935	3434.78	3373.55	江　苏
10971	74292	2091.19	2155.19	浙　江
12659	180976	4457.70	3666.20	安　徽
8608	64147	1971.82	1663.79	福　建
12636	157792	3138.89	2584.88	江　西
22136	88863	5309.68	4882.89	山　东
30268	183556	6709.28	5872.04	河　南
14690	115574	3320.27	2783.21	湖　北
17916	111052	4296.39	3448.80	湖　南
15260	145285	4682.31	3975.16	广　东
12779	167683	4105.96	3644.42	广　西
2083	18704	552.69	526.01	海　南
7131	56787	1894.43	1274.45	重　庆
22086	147424	5776.56	4439.38	四　川
10691	75135	2772.69	2277.37	贵　州
11841	130439	3434.38	3237.87	云　南
435	18003	240.79	236.64	西　藏
10876	68685	2145.20	1911.94	陕　西
8087	83687	1881.87	1683.23	甘　肃
1124	10074	367.88	350.55	青　海
1760	13277	371.98	307.40	宁　夏
5300	20940	1101.10	1056.98	新　疆
119	1756	84.94	58.09	新疆生产 建设兵团

3-2-26 村庄公共设施(一)(2020年)

地区名称 Name of Regions	集中供水的行政村 Administrative Villages With Access to Piped Water		年生活用 水　量 (万立方米) Annual Domestic Water Consumption (10,000 m^3)	供水管 道长度 (公里) Length of Water Supply Pipelines (km)	本年新增 Added This Year
	个数 (个) Number (unit)	比例 (%) Rate (%)			
全　国 National Total	**370423**	**82.48**	**1914067.08**	**1997955.83**	**142407.73**
北　京 Beijing	2777	87.19	17684.47	16668.58	77.06
天　津 Tianjin	2298	90.69	8177.10	12593.27	614.11
河　北 Hebei	34477	84.77	118169.55	165756.49	5256.65
山　西 Shanxi	16018	81.01	43929.28	62408.56	2232.17
内 蒙 古 Inner Mongolia	7416	72.58	22060.23	48565.56	2420.67
辽　宁 Liaoning	6257	64.49	35712.31	42893.07	3043.20
吉　林 Jilin	6843	82.15	25699.68	55457.59	9969.88
黑 龙 江 Heilongjiang	7154	90.66	33344.81	65015.22	1310.95
上　海 Shanghai	1267	91.94	16052.79	9587.86	93.31
江　苏 Jiangsu	12477	98.93	108770.94	106767.03	3161.84
浙　江 Zhejiang	11837	83.10	97874.32	62891.47	3646.48
安　徽 Anhui	11102	82.06	95187.50	83461.28	8442.30
福　建 Fujian	10676	92.59	56852.35	40221.26	2268.03
江　西 Jiangxi	10983	71.07	59386.14	49052.21	5088.12
山　东 Shandong	55403	97.02	153152.17	167388.71	9888.67
河　南 Henan	31215	80.69	145911.22	117638.82	9899.09
湖　北 Hubei	16686	81.41	84738.00	89983.76	6995.69
湖　南 Hunan	12471	61.26	73974.37	67435.92	4575.78
广　东 Guangdong	13615	81.41	129288.45	80005.12	6711.17
广　西 Guangxi	9220	68.90	98118.57	58654.32	6040.91
海　南 Hainan	2569	94.76	17408.31	10827.00	468.60
重　庆 Chongqing	6556	85.79	37642.09	37492.08	5158.49
四　川 Sichuan	19868	76.29	111027.54	114507.09	8539.07
贵　州 Guizhou	9302	72.87	67390.82	75199.04	8763.08
云　南 Yunnan	9956	82.06	98265.14	110093.80	11585.01
西　藏 Tibet	2251	47.15	10652.61	11801.01	829.29
陕　西 Shaanxi	13362	90.37	53376.18	54569.84	4105.16
甘　肃 Gansu	12402	83.85	40894.22	70275.79	4803.89
青　海 Qinghai	3082	78.38	9659.32	17543.87	274.09
宁　夏 Ningxia	1928	99.59	9415.34	18596.40	859.62
新　疆 Xinjiang	7556	92.53	32047.99	65906.81	4695.00
新疆生产 建设兵团 Xinjiang Production and Construction Corps	1399	86.68	2203.27	8697.00	590.35

Public Facilities of Villages Ⅰ (2020)

用水人口 （万人） Population with Access to Water (10,000 persons)	供水普及率 (%) Water Coverage Rate (%)	人均日生 活用水量 （升） Per Capita Daily Water Consumption (liter)	用气人口 （万人） Population with Access to Gas (10,000 persons)	燃气普及率 (%) Gas Coverage Rate (%)	集中供 热面积 （万平方米） Area of Centrally Heated District (10,000 m²)	地区名称 Name of Regions
56302.75	83.37	93.14	23690.02	35.08	32263.42	全　国
436.87	92.78	110.90	182.78	38.82	1184.85	北　京
222.58	93.25	100.65	169.41	70.97	1417.20	天　津
3914.85	92.26	82.70	2182.22	51.43	6298.71	河　北
1487.33	88.69	80.92	326.85	19.49	5428.32	山　西
749.91	71.80	80.59	115.49	11.06	1497.65	内　蒙　古
1101.71	69.90	88.81	232.78	14.77	1294.50	辽　宁
842.63	74.31	83.56	96.10	8.47	446.04	吉　林
1100.72	85.27	83.00	63.69	4.93	431.90	黑　龙　江
423.88	95.42	103.76	330.80	74.46	16.70	上　海
3290.63	97.54	90.56	2895.94	85.84	7.10	江　苏
1828.97	84.86	146.61	1096.35	50.87	9.90	浙　江
2903.72	79.20	89.81	1258.02	34.31	0.00	安　徽
1545.14	92.87	100.81	1048.74	63.03	0.00	福　建
1711.78	66.22	95.05	715.89	27.70	208.45	江　西
4667.49	95.59	89.90	2610.42	53.46	9099.64	山　东
4714.58	80.29	84.79	1252.42	21.33	723.88	河　南
2129.93	76.53	109.00	865.59	31.10	704.20	湖　北
2232.86	64.74	90.77	789.33	22.89	34.00	湖　南
3425.40	86.17	103.41	2629.40	66.15	0.00	广　东
2910.56	79.86	92.36	1923.86	52.79	23.00	广　西
479.74	91.20	99.42	386.50	73.48	0.00	海　南
1062.46	83.37	97.07	358.02	28.09	19.50	重　庆
3309.70	74.55	91.91	1414.01	31.85	0.00	四　川
1881.56	82.62	98.13	83.81	3.68	5.63	贵　州
2849.31	88.00	94.49	133.96	4.14	60.00	云　南
215.84	91.21	135.22	19.92	8.42	9.14	西　藏
1713.71	89.63	85.33	303.54	15.88	436.76	陕　西
1468.63	87.25	76.29	71.56	4.25	992.66	甘　肃
339.78	96.93	77.89	11.46	3.27	149.99	青　海
299.78	97.52	86.05	32.96	10.72	342.13	宁　夏
985.66	93.25	89.08	72.22	6.83	1252.50	新　疆
55.04	94.74	109.68	15.99	27.53	169.07	新疆生产 建设兵团

3-2-27　村庄公共设施(二)(2020年)

地区名称 Name of Regions	村 庄 内 道路长度 (公里) The Length of Roads within Villages (kilometer)	本年新增 Added This Year	本年更新 改　造 Renewal and Upgrading during The Reported Year	硬化道路 Hardened Roads
全　国　National Total	**3358041.38**	**105965.79**	**110688.21**	**1879943.74**
北　京　Beijing	18241.25	79.10	397.89	15538.41
天　津　Tianjin	16169.62	145.63	1035.20	9440.24
河　北　Hebei	217660.68	2117.53	7190.45	147249.30
山　西　Shanxi	81016.57	789.52	1526.08	41131.94
内　蒙　古　Inner Mongolia	84317.00	810.28	738.62	54739.32
辽　宁　Liaoning	75172.29	1301.44	4006.97	42741.99
吉　林　Jilin	84200.43	593.58	3935.24	58826.63
黑　龙　江　Heilongjiang	82367.87	187.06	1571.88	47556.34
上　海　Shanghai	11139.83	18.39	418.66	6785.03
江　苏　Jiangsu	143693.32	2955.19	4010.04	108447.93
浙　江　Zhejiang	80023.71	1757.56	2618.69	27432.08
安　徽　Anhui	157957.76	8793.83	4277.28	75632.05
福　建　Fujian	75224.13	1703.48	3231.88	43342.19
江　西　Jiangxi	97570.65	4696.85	3519.68	51513.55
山　东　Shandong	340607.85	9422.41	17956.98	231293.21
河　南　Henan	197900.89	10044.72	5627.88	91985.58
湖　北　Hubei	207323.96	6639.23	7231.44	75494.56
湖　南　Hunan	160259.16	4632.24	4346.47	66128.37
广　东　Guangdong	155013.23	7394.85	5641.86	78219.38
广　西　Guangxi	116092.49	4086.63	2573.24	72081.86
海　南　Hainan	28104.10	797.71	504.58	6185.08
重　庆　Chongqing	31355.85	2255.70	2389.61	15625.32
四　川　Sichuan	280641.46	12522.47	6520.81	196876.02
贵　州　Guizhou	127129.92	4741.93	3282.95	47736.09
云　南　Yunnan	148811.99	5568.43	5466.13	74233.63
西　藏　Tibet	18945.46	915.43	714.67	5556.46
陕　西　Shaanxi	100201.54	3125.42	3270.36	69031.18
甘　肃　Gansu	89423.10	2969.08	2696.25	50983.22
青　海　Qinghai	30633.88	695.80	360.57	13310.38
宁　夏　Ningxia	26117.73	787.62	1003.11	19684.88
新　疆　Xinjiang	65621.81	3240.17	2389.94	32058.89
新疆生产 建设兵团　Xinjiang Production and Construction Corps	9101.85	176.51	232.80	3082.63

Public Facilities of Villages Ⅱ (2020)

村庄内道路面积（万平方米）The Area of Roads within Villages (10,000 m²)	本年新增 Added This Year	本年更新改造 Renewal and Upgrading during The Reported Year	硬化道路 Hardened Roads	排水管道沟渠长度（公里）The Length of Drainage Pipelines and Canals (kilometer)	本年新增 Added This Year	地区名称 Name of Regions
2587761.82	117469.21	154836.81	1116257.50	1203114.45	52751.76	全　国
11026.08	67.65	278.56	8045.67	9614.77	198.24	北　京
9139.94	121.70	843.75	4281.35	6241.69	826.68	天　津
130475.41	1657.10	20384.65	75729.73	57755.82	1274.40	河　北
52321.04	831.89	1072.89	28263.24	56780.70	576.21	山　西
52621.65	717.25	594.55	31683.67	9463.94	137.50	内　蒙　古
42293.19	735.47	2428.63	23088.02	34721.21	387.83	辽　宁
43620.64	434.87	2398.78	27852.00	46968.27	671.86	吉　林
41358.11	138.17	829.06	21246.49	26819.75	375.96	黑　龙　江
18404.10	17.58	332.56	16192.33	6767.38	175.54	上　海
117560.62	7033.64	5313.96	83405.01	48605.04	2609.01	江　苏
77163.37	2688.07	2830.01	25640.72	40433.15	1237.19	浙　江
107298.40	9121.59	3686.83	39124.13	43438.30	2932.97	安　徽
43314.54	1481.22	1267.43	23353.78	61802.63	1942.08	福　建
99678.53	5040.21	2690.12	24264.08	39134.00	2908.12	江　西
218266.40	9488.57	12809.66	116955.36	228238.68	5475.46	山　东
352739.52	12395.47	45634.16	96305.21	55640.83	4567.48	河　南
198012.52	8186.23	9100.79	44878.63	58735.62	2896.11	湖　北
123584.75	4021.37	3176.28	34853.30	49227.42	1587.33	湖　南
130518.03	8637.34	6282.75	53287.08	59433.25	7746.19	广　东
65885.93	3859.82	2071.06	33809.46	28529.18	1369.04	广　西
17171.82	1155.02	463.57	3308.75	4321.13	350.79	海　南
16212.81	1518.12	1266.94	7860.65	13894.25	1050.33	重　庆
158605.79	9432.16	4718.05	91986.83	74804.42	3308.23	四　川
140314.00	5616.55	3331.36	47939.72	23775.85	1411.95	贵　州
103374.83	5982.47	4131.31	43239.52	42563.23	2430.03	云　南
24501.65	9268.12	10221.05	12150.85	2370.53	191.24	西　藏
51703.05	2195.47	2089.81	27997.98	29887.30	1466.73	陕　西
47537.53	1929.56	2139.56	24387.57	20096.60	1281.81	甘　肃
14350.56	409.26	223.21	6614.35	5063.34	111.41	青　海
12817.65	374.03	448.05	8829.45	10778.85	411.09	宁　夏
58148.65	2752.39	1593.11	26057.29	5598.81	607.66	新　疆
7740.71	160.85	184.31	3625.28	1608.51	235.29	新疆生产建设兵团

3-2-28 村庄房屋(2020年)

地区名称 Name of Regions	住宅 Residential Building						公共建筑	
	年末实有建筑面积(万平方米) Total Floor Space of Buildings (year-end) (10,000 m²)	混合结构以上 Mixed Strucure and Above	本年竣工建筑面积(万平方米) Floor Space Completed This Year (10,000 m²)	混合结构以上 Mixed Strucure and Above	房地产开发 Real Estate Development	人均住宅建筑面积(平方米) Per Capita Floor Space (m²)	年末实有建筑面积(万平方米) Total Floor Space of Buildings (year-end) (10,000 m²)	混合结构以上 Mixed Strucure and Above
全 国 National Total	2664595.91	2048138.37	75566.27	64771.23	4317.04	34.31	166097.27	135185.84
北 京 Beijing	16068.25	13899.26	1381.40	214.54	0.00	49.00	3320.94	3278.11
天 津 Tianjin	7458.04	6192.75	46.19	42.11	3.33	31.42	492.37	473.21
河 北 Hebei	132538.01	103495.31	1444.13	1030.16	168.05	28.32	8534.22	7352.62
山 西 Shanxi	58831.46	38205.98	1080.12	650.19	160.68	30.25	5583.56	4099.81
内 蒙 古 Inner Mongolia	36291.87	31802.44	354.30	284.18	155.20	27.11	2259.74	2061.88
辽 宁 Liaoning	40530.23	27035.97	282.71	259.45	19.01	23.66	2433.43	2047.94
吉 林 Jilin	31298.11	23746.91	145.29	128.25	7.08	23.78	1192.25	1008.05
黑 龙 江 Heilongjiang	39351.03	35260.29	212.74	155.60	0.00	23.61	1554.25	1517.63
上 海 Shanghai	12027.10	10865.28	19.32	12.32	8.00	39.60	780.93	653.11
江 苏 Jiangsu	145237.65	122584.19	1691.98	1628.19	128.88	42.28	8154.58	7297.94
浙 江 Zhejiang	129229.67	110542.33	1777.50	1631.12	173.70	61.80	7578.14	6771.82
安 徽 Anhui	148167.85	111584.12	2889.30	2383.72	252.54	33.24	7627.58	4783.39
福 建 Fujian	75391.59	55654.70	1592.59	1417.54	176.05	38.23	6948.28	5598.27
江 西 Jiangxi	120937.97	102232.04	4077.57	3863.55	47.34	38.53	8706.66	7363.92
山 东 Shandong	170315.16	128071.68	3755.82	2706.82	709.96	32.08	14500.09	13169.60
河 南 Henan	210070.95	165742.22	6226.38	4879.67	524.24	31.31	13744.40	11704.02
湖 北 Hubei	115872.95	86964.95	3407.53	2533.26	109.40	34.90	8496.83	5593.60
湖 南 Hunan	133461.87	108129.58	4106.41	3149.48	206.72	31.06	7932.21	7112.37
广 东 Guangdong	159563.07	131828.29	7078.48	6295.07	649.28	34.08	8168.18	6541.45
广 西 Guangxi	120255.74	91191.97	2642.96	2533.68	78.95	29.29	6478.08	5592.44
海 南 Hainan	17141.55	16918.35	301.35	202.72	3.69	31.01	723.97	712.52
重 庆 Chongqing	73943.72	59140.06	1345.38	1271.67	6.24	39.03	4176.10	3761.94
四 川 Sichuan	214843.22	168313.43	3516.46	3142.42	197.28	37.19	8109.36	6322.80
贵 州 Guizhou	105023.35	58563.52	6454.82	5706.17	25.61	37.88	4569.32	3338.02
云 南 Yunnan	128302.00	81196.01	4038.58	3583.00	345.50	37.36	9521.77	5539.70
西 藏 Tibet	8518.32	3369.89	4255.56	4200.82	0.74	35.38	2482.07	983.12
陕 西 Shaanxi	98103.00	72308.53	834.98	736.99	40.85	45.73	3752.85	3346.68
甘 肃 Gansu	52824.83	37237.71	960.18	788.48	67.38	28.07	3916.30	3293.88
青 海 Qinghai	10951.09	7004.84	76.36	55.72	0.00	29.77	592.84	490.12
宁 夏 Ningxia	11470.13	8650.29	221.88	196.85	0.54	30.84	688.25	644.74
新 疆 Xinjiang	38873.34	29507.30	9287.68	9051.24	50.79	35.30	2893.15	2607.12
新疆生产建设兵团 Xinjiang Production and Construction Corps	1702.78	898.18	60.32	36.24	0.00	20.05	184.60	123.99

Building Construction of Villages(2020)

Public Building			生产性建筑 Industrial Building					
本年竣工建筑面积(万平方米)Floor Space Completed This Year (10,000 m²)	混合结构以上 Mixed Strucure and Above	房地产开发 Real Estate Development	年末实有建筑面积(万平方米)Total Floor Space of Buildings (year-end) (10,000 m²)	混合结构以上 Mixed Strucure and Above	本年竣工建筑面积(万平方米)Floor Space Completed This Year (10,000 m²)	混合结构以上 Mixed Strucure and Above	房地产开发 Real Estate Development	地区名称 Name of Regions
10715.59	**8012.82**	**446.39**	**244038.64**	**187656.62**	**10034.03**	**8038.18**	**594.59**	全　国
64.53	64.43	0.00	2341.66	2271.36	0.46	0.39	0.00	北　京
24.20	23.38	0.00	2302.62	1998.01	11.09	7.69	0.00	天　津
263.24	181.38	2.58	15661.04	11867.77	251.16	175.87	2.86	河　北
1027.42	1016.02	3.23	5575.00	3992.22	114.39	89.30	1.00	山　西
33.36	28.79	0.65	2685.36	2168.38	94.98	79.06	1.22	内　蒙古
23.18	17.67	4.17	4787.21	3593.46	122.00	95.25	0.21	辽　宁
46.06	44.37	0.00	1685.60	1498.68	49.71	48.09	0.14	吉　林
6.02	5.78	0.10	2600.67	2396.78	81.67	68.23	0.00	黑龙江
15.60	10.50	0.00	4412.51	4033.38	24.52	20.92	0.00	上　海
264.77	254.29	14.45	26892.70	24176.65	698.92	649.32	9.47	江　苏
385.18	364.54	20.49	22006.26	19577.40	694.64	653.19	0.40	浙　江
333.36	286.78	13.89	7013.61	4897.19	375.77	308.99	25.64	安　徽
261.62	208.36	10.94	15257.57	8777.28	355.00	294.47	84.48	福　建
1152.60	846.24	18.56	6868.97	5407.70	843.15	504.95	4.50	江　西
524.71	489.62	63.93	27310.82	21882.11	1088.95	942.42	115.81	山　东
605.03	469.76	92.62	16280.20	13094.38	639.58	539.48	57.66	河　南
506.56	308.42	19.53	6986.72	5099.22	346.04	287.94	24.57	湖　北
457.54	412.73	20.37	7942.53	7103.78	414.25	311.34	19.15	湖　南
834.40	516.96	97.44	13459.50	8625.94	1061.49	777.41	178.50	广　东
224.24	200.78	1.79	3421.60	2235.83	218.66	116.61	0.28	广　西
21.10	20.83	0.85	308.57	293.36	11.51	9.57	0.92	海　南
47.39	45.38	0.10	4756.31	3570.11	107.95	91.43	0.16	重　庆
292.72	258.97	13.51	9792.96	7883.02	511.25	398.03	10.97	四　川
312.38	278.51	4.57	5003.75	3050.87	324.02	249.26	4.17	贵　州
609.11	572.85	22.29	10649.33	7069.71	481.21	352.47	22.30	云　南
1805.61	579.76	1.61	715.77	641.59	25.13	21.95	0.20	西　藏
164.17	148.30	9.13	3598.27	2860.61	177.02	158.49	0.60	陕　西
163.88	131.77	3.62	7454.61	3020.64	186.02	158.84	0.14	甘　肃
15.87	12.10	0.00	510.60	331.61	18.89	16.32	0.04	青　海
28.56	25.28	0.00	1255.76	969.30	72.08	54.79	0.00	宁　夏
189.78	179.10	5.98	4038.62	2980.93	609.11	535.65	28.37	新　疆
11.38	9.14	0.00	461.89	287.36	23.43	20.45	0.84	新疆生产建设兵团

3-2-29 村庄建设投入(2020年)

计量单位: 万元

地区名称 Name of Regions	本年建设投入合计 Total Construction Input of This Year	房屋 Building					小计 Input of Municipal Public Facilities	供水 Water Supply
		小计 Input of House	房地产开发 Real Estate Development	住宅 Residential Building	公共建筑 Public Building	生产性建筑 Industrial Building		
全 国 **National Total**	**115034156**	**79133431**	**12240545**	**56696884**	**8559286**	**13877262**	**35900725**	**4022320**
北 京 Beijing	756198	327131	3920	253964	70828	2339	429067	28083
天 津 Tianjin	492572	110116	1590	51011	43590	15515	382456	80366
河 北 Hebei	3893362	2734970	310261	1697342	613711	423916	1158393	78430
山 西 Shanxi	2141388	1596587	381154	1337818	137087	121682	544801	40744
内 蒙 古 Inner Mongolia	851374	594646	11680	450882	45635	98129	256727	39370
辽 宁 Liaoning	779011	447020	16762	266581	22962	157477	331991	33505
吉 林 Jilin	848373	275696	1214	159674	53768	62253	572676	159212
黑 龙 江 Heilongjiang	463353	221351	14233	161138	7171	53042	242002	24121
上 海 Shanghai	1019895	187977	54933	35530	151440	1006	831919	6010
江 苏 Jiangsu	6061322	3952015	504371	2639791	368643	943581	2109307	161807
浙 江 Zhejiang	7039633	4943839	1877267	3551946	442051	949841	2095794	287879
安 徽 Anhui	5516686	3749772	997806	3003276	360701	385796	1766914	226838
福 建 Fujian	4028428	3008877	858189	2259615	316832	432430	1019551	90445
江 西 Jiangxi	4286789	3181700	153934	2453207	511968	216524	1105090	118614
山 东 Shandong	11107370	7068441	1893239	4977096	732718	1358627	4038929	382838
河 南 Henan	7358204	5707536	745421	4584536	521304	601696	1650668	174386
湖 北 Hubei	3330097	2083459	152641	1398896	353028	331535	1246639	167417
湖 南 Hunan	5572777	3580404	213294	2724105	411177	445122	1992373	176052
广 东 Guangdong	10607935	8081537	2269366	6355143	699650	1026745	2526397	291400
广 西 Guangxi	3769836	2556060	110783	2185215	197695	173150	1213776	196932
海 南 Hainan	476698	295051	2977	262932	23115	9004	181647	15918
重 庆 Chongqing	2061246	1120394	128640	944851	70537	105006	940852	81899
四 川 Sichuan	8037609	5147554	384269	4122710	391134	633710	2890056	198126
贵 州 Guizhou	4119633	2518548	127928	1988511	293988	236048	1601086	210143
云 南 Yunnan	9130698	7030213	637054	5249542	871291	909381	2100485	267910
西 藏 Tibet	3754811	3528247	19524	165394	191222	3171631	226564	19917
陕 西 Shaanxi	2021491	1264596	41400	891554	173482	199561	756895	154354
甘 肃 Gansu	1942356	1412343	136625	1046101	159716	206526	530013	75696
青 海 Qinghai	208830	131462	2777	91090	22102	18270	77368	6285
宁 夏 Ningxia	574403	382398	30951	263388	46058	72952	192006	17290
新 疆 Xinjiang	2186580	1528698	148149	1096118	231293	201287	657882	167086
新疆生产建设兵团 Xinjiang Production and Construction Corps	595198	364795	8195	27927	23388	313479	230403	43246

Construction Input of Villages(2020)

Measurement Unit: 10,000 RMB

燃 气 Gas Supply	集中供热 Central Heating	道路桥梁 Road and Bridge	排 水 Drainage	市政公用设施 Municipal Public Facilities			其 他 Other	地区名称 Name of Regions
				污水处理 Wastewater Treatment	园林绿化 Landscaping	环境卫生 Environmental Sanitation	垃圾处理 Garbage Treatment	其 他 Other
1806222	355332	15560876	5439444	3524426	2050536	4803526	2776661	1862469	全　国
2227	5851	53066	62186	35373	167012	103275	30809	7367	北　京
29107	15488	52859	145581	56321	17174	34349	13117	7531	天　津
428139	54413	304039	78068	45494	55691	143389	82276	16223	河　北
73172	65163	212855	34479	11403	29416	77133	24996	11839	山　西
6859	11505	94325	33955	13669	22183	37155	23152	11376	内 蒙 古
3640	9171	167827	34230	25014	11867	62139	40153	9613	辽　宁
1275	822	245995	44755	18298	20595	81260	37470	18762	吉　林
3077	416	101831	22527	14715	8581	55150	34846	26299	黑 龙 江
11608		129080	457398	354616	41757	50248	19004	135819	上　海
52933	520	617139	572214	427025	201927	358477	152370	144289	江　苏
32943	656	791638	328694	204994	181045	259438	132962	213502	浙　江
24436	711	897504	232816	130068	99011	210180	114043	75419	安　徽
11935	10	391976	238609	169074	95619	132085	86344	58871	福　建
9640	431	491036	162271	70771	74671	138779	75444	109649	江　西
542581	122311	1664111	354701	212178	335447	445868	219291	191072	山　东
248515	11722	590093	234041	125515	153696	188647	95496	49569	河　南
25822	163	520071	188018	93169	97786	142255	80626	105107	湖　北
16043	192	1403171	135512	67827	52967	131687	79466	76749	湖　南
22689	28010	758515	1015698	842757	80874	233383	126923	95828	广　东
4224	44	769672	98161	39403	15126	86136	62096	43481	广　西
5433		72028	50777	33808	6863	24814	10991	5813	海　南
28408	24	525887	90499	66385	23371	124925	99120	65840	重　庆
114163		2068797	233798	150489	40427	135875	79923	98869	四　川
15648	185	797689	112565	57476	18896	396731	242859	49229	贵　州
3033	100	696246	169882	115370	55603	850208	669430	57504	云　南
439	445	180074	8973	4126	5121	5771	4690	5824	西　藏
64264	5445	264069	73140	19513	66587	92293	43290	36743	陕　西
1620	6415	220194	55157	29552	18513	65298	37035	87119	甘　肃
1474	15	36491	11782	7332	3485	13358	6275	4478	青　海
2080	1389	63414	44305	34665	13081	30802	10961	19644	宁　夏
15621	13195	270513	82048	37958	21383	75092	35876	12944	新　疆
3176	520	108670	32604	10066	14762	17329	5329	10096	新疆生产 建设兵团

主要指标解释

Explanatory Notes on Main Indicators

主要指标解释

城市和县城部分

人口密度

指城区内的人口疏密程度。计算公式：

$$人口密度 = \frac{城区人口 + 城区暂住人口}{城区面积}$$

人均日生活用水量

指每一用水人口平均每天的生活用水量。计算公式：

$$人均日生活用水量 = \frac{居民家庭用水量 + 公共服务用水量 + 免费供水量中的生活用水量}{用水人口} \div 报告期日历天数 \times 1000升$$

供水普及率

指报告期末城区内用水人口与总人口的比率。计算公式：$人口密度 = \dfrac{城区人口 + 城区暂住人口}{城区面积}$

$$供水普及率 = \frac{城区用水人口（含暂住人口）}{城区人口 + 城区暂住人口}$$

$$公共供水普及率 = \frac{城区公共用水人口（含暂住人口）}{城区人口 + 城区暂住人口} \times 100\%$$

燃气普及率

指报告期末城区内使用燃气的人口与总人口的比率。计算公式：

$$燃气普及率 = \frac{城区用气人口（含暂住人口）}{城区人口 + 城区暂住人口} \times 100\%$$

人均城市道路面积

指报告期末城区内平均每人拥有的道路面积。计算公式：

$$人均道路面积 = \frac{城区道路面积}{城区人口 + 城区暂住人口}$$

建成区路网密度

指报告期末建成区内道路分布的稀疏程度。计算公式：

$$建成区路网密度 = \frac{建成区道路长度}{建成区面积}$$

建成区排水管道密度

指报告期末建成区排水管道分布的疏密程度。计算公式：

$$建成区排水管道密度 = \frac{建成区排水管道长度}{建成区面积}$$

污水处理率

指报告期内污水处理总量与污水排放总量的比率。计算公式：

$$污水处理率 = \frac{污水处理总量}{污水排放总量} \times 100\%$$

污水处理厂集中处理率

指报告期内通过污水处理厂处理的污水量与污水排放总量的比率。计算公式：

$$污水处理厂集中处理率=\frac{污水处理厂处理的污水量}{污水排放总量}\times100\%$$

人均公园绿地面积

指报告期末城区内平均每人拥有的公园绿地面积。计算公式：

$$人均公园绿地面积=\frac{城区公园绿地面积}{城区人口+城区暂住人口}$$

建成区绿化覆盖率

指报告期末建成区内绿化覆盖面积与区域面积的比率。计算公式：

$$建成区绿化覆盖率=\frac{建成区绿化覆盖面积}{建成区面积}\times100\%$$

建成区绿地率

指报告期末建成区内绿地面积与建成区面积的比率。计算公式：

$$建成区绿地率=\frac{建成区绿地面积}{建成区面积}\times100\%$$

生活垃圾处理率

指报告期内生活垃圾处理量与生活垃圾产生量的比率。计算公式：

$$生活垃圾处理率=\frac{生活垃圾处理量}{生活垃圾产生量}\times100\%$$

生活垃圾无害化处理率

指报告期内生活垃圾无害化处理量与生活垃圾产生量的比率。计算公式：

$$生活垃圾无害化处理率=\frac{生活垃圾无害化处理量}{生活垃圾产生量}\times100\%$$

在统计时，由于生活垃圾产生量不易取得，可用清运量代替。"垃圾清运量"在审核时要与总人口（包括暂住人口）对应，一般城市人均日产生垃圾为 1kg 左右。

市区（县）面积

指城市（县）行政区域内的全部土地面积(包括水域面积)。地级以上城市行政区不包括市辖县(市)。按国务院批准的行政区划面积为准填报。

城区（县城）面积

指城市的城区和县城的面积。

设市城市城区包括：市本级(1)街道办事处所辖地域；(2)城市公共设施、居住设施和市政公用设施等连接到的其他镇（乡）地域；（3）常住人口在 3000 人以上独立的工矿区、开发区、科研单位、大专院校等特殊区域。

县城包括：(1)县政府驻地的镇（城关镇）或街道办事处地域；(2)县城公共设施、居住设施和市政设施等连接到的其他镇（乡）地域；（3）县域内常住人口在 3000 人以上独立的工矿区、开发区、科研单位、大专院校等特殊区域。

连接是指两个区域间可观察到的已建成或在建的公共设施、居住设施、市政设施和其他设施相连，中间没有被水域、农业用地、园地、林地、牧草地等非建设用地隔断。

对于组团式和散点式的城市，城区由多个分散的区域组成，或有个别区域远离主城区，应将这些分散的区域相加作为城区。

在统计时，以镇（乡）一级为最小统计划分单位，原则上不要打破镇（乡）的行政区划。

城区（县城）人口

指划定的城区（县城）范围的户籍人口数。按公安部门的统计为准填报。

暂住人口

指离开常住户口地的市区或乡、镇，到本市居住半年以上的人员。按市区、县和城区、县城分别统计，一般按公安部门的暂住人口统计为准填报。

建成区面积

指城区（县城）内实际已成片开发建设、市政公用设施和公共设施基本具备的区域。对核心城市，它包括集中连片的部分以及分散的若干个已经成片建设起来，市政公用设施和公共设施基本具备的地区；对一城多镇来说，它包括由几个连片开发建设起来的，市政公用设施和公共设施基本具备的地区组成。因此建成区范围，一般是指建成区外轮廓线所能包括的地区，也就是这个城市实际建设用地所达到的范围。

城市建设用地面积

指城市内的居住用地、公共管理与公共服务用地、商业服务业设施用地、工业用地、物流仓储用地、道路与交通设施用地、公用设施用地、绿地与广场用地。分别统计规划建设用地和现状建设用地。

规划建设用地面积指截至报告期末对应有关城市、县人民政府所在地镇的最新版依法批准的城市总体规划确定的城市（镇）用地面积。

现状建设用地面积指报告期末对应有关城市、县人民政府所在地镇的城市建设用地实际情况的面积。

城市维护建设资金

指用于城市维护和建设的资金，资金来源包括城市维护建设税、公用事业附加、中央和地方财政拨款、国内贷款、债券收入、利用外资、土地出让转让收入、资产置换收入、市政公用企事业单位自筹资金、国家和省规定收取的用于城市维护建设的行政事业性收费、集资收入以及其他收入。资金支出包括固定资产投资支出、维护支出和其他支出。

本年鉴中仅统计了用于城市维护和建设的财政性资金。

城市维护建设税

指依据《中华人民共和国城市维护建设税暂行条例》开征的一种地方性税种。现行的征收办法是以纳税人实际缴纳的增值税、消费税、营业税税额为计税依据，与增值税、消费税、营业税同时缴纳。根据纳税人所在地不同执行不同的纳税率：市区的税率为百分之七；县城、镇的税率为百分之五；不在市区、县城或镇的税率为百分之一。

城市公用事业附加

指在部分公用事业产品（服务）价外征收的用于城市维护建设的附加收入。包括工业用电、工业用水附加，公共汽车、电车、民用自来水、民用照明用电、电话、煤气、轮渡等附加。

固定资产投资

指建造和购置市政公用设施的经济活动，即市政公用设施固定资产再生产活动。市政公用设施固定资产再生产过程包括固定资产更新（局部更新和全部更新）、改建、扩建、新建等活动。新的企业财务会计制度规定，固定资产局部更新的大修理作为日常生产活动的一部分，发生的大修理费用直接在成本费用中列支。按照现行投资管理体制及有关部门的规定，凡属于养护、维护性质的工程，不纳入固定资产投资统计。对新建和对现有市政公用设施改造工程，应纳入固定资产统计。

本年新增固定资产

指在报告期已经完成建造和购置过程，并交付生产或使用单位的固定资产价值。包括已经建成投入生产或交付使用的工程投资和达到固定资产标准的设备、工具、器具的投资及有关应摊入的费用。属于增加固定资产价值的其他建设费用，应随同交付使用的工程一并计入新增固定资产。

新增生产能力（或效益）

指通过固定资产投资活动而增加的设计能力。计算新增生产能力(或效益)是以能独立发挥生产能力(或效益)的工程为对象。当工程建成,经有关部门验收鉴定合格,正式移交投入生产,即应计算新增生产能力(或效益)。

综合生产能力

指按供水设施取水、净化、送水、出厂输水干管等环节设计能力计算的综合生产能力。包括在原设计能力的基础上,经挖、革、改增加的生产能力。计算时,以四个环节中最薄弱的环节为主确定能力。对于经过更新改造,按更新改造后新的设计能力填报。

供水管道长度

指从送水泵至各类用户引入管之间所有市政管道的长度。不包括新安装尚未使用、水厂内以及用户建筑物内的管道。在同一条街道埋设两条或两条以上管道时,应按每条管道的长度计算。

供水总量

指报告期供水企业(单位)供出的全部水量。包括有效供水量和漏损水量。

有效供水量指水厂将水供出厂外后,各类用户实际使用到的水量。包括售水量和免费供水量。

漏损水量

指在供水工程中由于管道及附属设施破损而造成的漏失水量,以及计量损失量和其他损失水量。计量损失量包括居民用户总分表差损失水量、非居用户表具误差损失水量;其他损失水量是指未注册用户用水和用户拒查等管理因素导致的损失水量。

新水取用量

指取自任何水源被第一次利用的水量,包括自来水、地下水、地表水。新水量就一个城市来说,包括城市供水企业新水量和社会各单位的新水量。

其中:**工业新水取用量**指为使工业生产正常进行,保证生产过程对水的需要,而实际从各种水源引取的、为任何目的所用的新鲜水量,包括间接冷却水新水量、工艺水新水量、锅炉新水量及其他新水量。

用水重复利用量

指各用水单位在生产和生活中,循环利用的水量和直接或经过处理后回收再利用的水量之和。

其中:**工业用水重复利用量**指工业企业内部生活及生产用水中,循环利用的水量和直接或经过处理后回收再利用的水量之和。

节约用水量

指报告期新节水量,通过采用各项节水措施(如改进生产工艺、技术、生产设备、用水方式、换装节水器具、加强管理等)后,用水量和用水效益产生效果,而节约的水量。

人工煤气生产能力

指报告期末燃气生产厂制气、净化、输送等环节的综合生产能力,不包括备用设备能力。一般按设计能力计算,如果实际生产能力大于设计能力时,应按实际测定的生产能力计算。测定时应以制气、净化、输送三个环节中最薄弱的环节为主。

供气管道长度

指报告期末从气源厂压缩机的出口或门站出口至各类用户引入管之间的全部已经通气投入使用的管道长度。不包括煤气生产厂、输配站、液化气储存站、灌瓶站、储配站、气化站、混气站、供应站等厂(站)内,以及用户建筑物内的管道。

供气总量

指报告期燃气企业(单位)向用户供应的燃气数量。包括销售量和损失量。

汽车加气站

指专门为燃气机动车(船舶)提供压缩天然气、液化石油气等燃料加气服务的站点。应按不同气种分别统计。

供热能力

指供热企业(单位)向城市热用户输送热能的设计能力。

供热总量

指在报告期供热企业（单位）向城市热用户输送全部蒸汽和热水的总热量。

供热管道长度

指从各类热源到热用户建筑物接入口之间的全部蒸汽和热水的管道长度。不包括各类热源厂内部的管道长度。

其中：一级管网指由热源至热力站间的供热管道，**二级管网**指热力站至用户之间的供热管道。

城市道路

指城市供车辆、行人通行的，具备一定技术条件的道路、桥梁、隧道及其附属设施。城市道路由车行道和人行道等组成。在统计时只统计路面宽度在 3.5 米（含 3.5 米）以上的各种铺装道路，包括开放型工业区和住宅区道路在内。

道路长度

指道路长度和与道路相通的桥梁、隧道的长度，按车行道中心线计算。

道路面积

指道路面积和与道路相通的广场、桥梁、隧道的铺装面积（统计时，将车行道面积、人行道面积分别统计）。

人行道面积按道路两侧面积相加计算，包括步行街和广场，不含人车混行的道路。

桥梁

指为跨越天然或人工障碍物而修建的构筑物。包括跨河桥、立交桥、人行天桥以及人行地下通道等。

道路照明灯盏数

指在城市道路设置的各种照明用灯。一根电杆上有几盏即计算几盏。统计时，仅统计功能照明灯，不统计景观照明灯。

防洪堤长度

指实际修筑的防洪堤长度。统计时应按河道两岸的防洪堤相加计算长度，但如河岸一侧有数道防洪堤时，只计算最长一道的长度。

污水排放总量

指生活污水、工业废水的排放总量，包括从排水管道和排水沟（渠）排出的污水量。

（1）可按每条管道、沟（渠）排放口的实际观测的日平均流量与报告期日历日数的乘积计算。

（2）有排水测量设备的，可按实际测量值计算。

（3）如无观测值，也可按当地供水总量乘以污水排放系数确定。

城市分类污水排放系数

城市污水分类	污水排放系数
城市污水	0.7～0.8
城市综合生活污水	0.8～0.9
城市工业废水	0.7～0.9

排水管道长度

指所有市政排水总管、干管、支管、检查井及连接井进出口等长度之和。计算时应按单管计算，即在同一条街道上如有两条或两条以上并排的排水管道时，应按每条排水管道的长度相加计算。

其中：**污水管道**指专门排放污水的排水管道。

雨水管道指专门排放雨水的排水管道。

雨污合流管道指雨水、污水同时进入同一管道进行排水的排水管道。

污水处理量

指污水处理厂（或污水处理装置）实际处理的污水量。包括物理处理量、生物处理量和化学处理量。

其中**处理本城区（县城）外**，指污水处理厂作为区域设施，不仅处理本城区（县城）的污水，还处理本市（县）以外其他市、县或本市（县）其他乡村等的污水。这部分污水处理量单独统计，并在计算本市（县）的污水处理率时扣除。

干污泥年产生量

指全年污水处理厂在污水处理过程中干污泥的最终产生量。干污泥是指以干固体质量计的污泥量，含水率为0。如果产生的湿污泥的含水率为n%，那么干污泥产生量=湿污泥产生量×（1-n%）。

干污泥处置量

指报告期内将污泥达标处理处置的干污泥量。统计时按土地利用、建材利用、焚烧、填埋和其他分别填写。其中：

污泥土地利用指处理达标后的污泥产物用于园林绿化、土地改良、林地、农用等场合的处置方式。

污泥建筑材料利用指将污泥处理达标后的产物作为制砖、水泥熟料等建筑材料部分原料的处置方式。

污泥焚烧指利用焚烧炉将污泥完全矿化为少量灰烬的处理处置方式，包括单独焚烧，以及与生活垃圾、热电厂等工业窑炉的协同焚烧。

污泥填埋指采取工程措施将处理达标后的污泥产物进行堆、填、埋，置于受控制场地内的处置方式。

绿化覆盖面积

指城市中乔木、灌木、草坪等所有植被的垂直投影面积。包括城市各类绿地绿化种植垂直投影面积、屋顶绿化植物的垂直投影面积以及零星树木的垂直投影面积，乔木树冠下的灌木和草本植物以及灌木树冠下的草本植物垂直投影面积均不能重复计算。

绿地面积

指报告期末用作园林和绿化的各种绿地面积。包括公园绿地、防护绿地、广场用地、附属绿地和位于建成区范围内的区域绿地面积。

其中：**公园绿地**指向公众开放，以游憩为主要功能，兼具生态、景观、文教和应急避险等功能，有一定游憩和服务设施的绿地。

防护绿地指用地独立，具有卫生、隔离、安全、生态防护功能，游人不宜进入的绿地。主要包括卫生隔离防护绿地、道路及铁路防护绿地、高压走廊防护绿地、公共设施防护绿地等。

广场用地指以游憩、纪念、集会和避险等功能为主的城市公共活动场地。

附属绿地指附属于各类城市建设用地（除"绿地与广场用地"）的绿化用地。包括居住用地、公共管理与公共服务设施用地、商业服务业设施用地、工业用地、物流仓储用地、道路和交通设施用地、公共设施用地等用地中的绿地。

区域绿地指位于城市建设用地之外，具有城乡生态环境及自然资源和文化资源保护、游憩健身、安全防护隔离、物种保护、园林苗木生产等功能的绿地。

公园

指常年开放的供公众游览、观赏、休憩、开展科学、文化及休闲等活动，有较完善的设施和良好的绿化环境、景观优美的公园绿地。包括综合性公园、儿童公园、文物古迹公园、纪念性公园、风景名胜公园、动物园、植物园、带状公园等。不包括居住小区及小区以下的游园。统计时只统计市级和区级的综合公园、专类公园和带状公园。

其中：**门票免费公园**指对公众免费开放，不售门票的公园。

道路清扫保洁面积

指报告期末对城市道路和公共场所（主要包括城市行车道、人行道、车行隧道、人行过街地下通道、道路附属绿地、地铁站、高架路、人行过街天桥、立交桥、广场、停车场及其他设施等）进行清扫保洁的面积。一天清扫保洁多次的，按清扫保洁面积最大的一次计算。

其中：**机械化道路清扫保洁面积**指报告期末使用扫路车（机）、冲洗车等大小型机械清扫保洁的道路面积。多种机械在一条道路上重复使用时，只按一种机械清扫保洁的面积计算，不能重复统计。

生活垃圾、建筑清运量

指报告期收集和运送到各生活垃圾、建筑垃圾厂和生活垃圾、建筑垃圾最终消纳点的生活垃圾、建筑垃圾的数量。统计时仅计算从生活垃圾、建筑垃圾源头和从生活垃圾转运站直接送到处理场和最终消纳点的清运量，对于二次中转的清运量不要重复计算。

餐厨垃圾属于生活垃圾的一部分，无论单独清运还是混合清运，都应统计在生活垃圾清运量中。

其中：**餐厨垃圾清运处置量**指单独清运，并且进行单独处置的餐厨垃圾总量，不含混在生活垃圾中清运的部分。

公共厕所

指供城市居民和流动人口使用，在道路两旁或公共场所等处设置的厕所。分为独立式、附属式和活动式三种类型。统计时只统计独立式和活动式，不统计附属式公厕。

独立式公共厕所按建筑类别应分为三类，活动式公共厕所按其结构特点和服务对象应分为组装厕所、单体厕所、汽车厕所、拖动厕所和无障碍厕所五种类别。

市容环卫专用车辆设备

指用于环境卫生作业、监察的专用车辆和设备，包括用于道路清扫、冲洗、洒水、除雪、垃圾粪便清运、市容监察以及与其配套使用的车辆和设备。如：垃圾车、扫路机（车）、洗路车、洒水车、真空吸粪车、除雪机、装载机、推土机、压实机、垃圾破碎机、垃圾筛选机、盐粉撒布机、吸泥渣车和专用船舶等。对于长期租赁的车辆及设备也统计在内。

统计时，单独统计道路清扫保洁专用车辆和生活垃圾运输专用车辆数。

村镇部分

人口密度

指建成区范围内的人口疏密程度。计算公式：

$$建成区人口密度（人/平方公里）＝\frac{建成区常住人口（人）}{建成区面积（公顷）}×100$$

供水普及率

指报告期末建成区（村庄）用水人口与建成区（村庄）人口的比率。按建成区、村庄分别统计。计算公式：

$$建成区供水普及率（\%）＝\frac{建成区用水人口（人）}{建成区常住人口（人）}×100\%$$

$$村庄供水普及率（\%）＝\frac{村庄用水人口（人）}{村庄常住人口（人）}×100\%$$

燃气普及率

指报告期末建成区使用燃气的人口与建成区人口的比率。计算公式：

$$燃气普及率（\%）＝\frac{建成区用气人口（人）}{建成区常住人口（人）}×100\%$$

人均道路面积

指报告期末建成区范围内平均每人拥有的道路面积。计算公式：

$$建成区人均道路面积（平方米／人）＝\frac{建成区道路面积（万平方米）}{建成区常住人口（人）}\times10^4$$

排水管道暗渠密度

指报告期末建成区范围内排水管道暗渠分布的疏密程度。计算公式：

$$建成区排水管道暗渠密度（公里/平方公里）＝\frac{建成区排水管道长度（公里）+建成区排水暗渠长度（公里）}{建成区面积（公顷）}\times100$$

人均公园绿地面积

指报告期末建成区范围内平均每人拥有的公园绿地面积。计算公式：

$$建成区人均公园绿地面积（平方米／人）＝\frac{建成区公园绿地面积（公顷）}{建成区常住人口（人）}\times10^4$$

建成区绿化覆盖率

指报告期末建成区范围内绿化覆盖面积与建成区面积的比率。计算公式：

$$建成区绿化覆盖率 (\%)＝\frac{建成区绿化覆盖面积（公顷）}{建成区面积（公顷）}\times100\%$$

建成区绿地率

指报告期末镇（乡）建成区范围内绿地面积与建成区面积的比率。计算公式：

$$建成区绿地率 (\%)＝\frac{建成区绿地面积（公顷）}{建成区面积（公顷）}\times100\%$$

人均日生活用水量

指用水人口平均每天的生活用水量。计算公式：

人均日生活用水量=报告期生活用水量/用水人口/报告期日历天数×1000升

建成区

指行政区域内实际已成片开发建设、市政公用设施和公共设施基本具备的区域。建成区范围一般是指建成区外轮廓线所包括的地区，也就是实际建设用地达到的范围。建成区面积以镇人民政府建设部门(或规划部门)提供的范围为准。

村庄

指农村居民生活和生产的聚居点。

因为镇（乡）政府驻地和村庄的市政公用设施差距较大，所以在报表中，我们将镇、乡、农场等特殊区域分成建成区和村庄两部分进行统计。

村庄规划

指在镇总体规划或乡规划等指导下，根据经济发展水平，主要对住宅和供水、供电、道路、绿化、环境卫生以及生产配套设施建设的具体安排。村庄建设规划须经村民委员会议讨论同意，并由上级批准同意实施。统计时要注意规划的有效期，如至本年年底，规划已过期，则视为无规划。

本年村镇规划编制投入

指本年镇（乡）域内总体规划、专项规划、详细规划、村庄规划等各种规划编制工作的总投入。

公共供水

指公共供水企业以公共供水管道及其附属设施向单位和居民的生活、生产和其他各项建设提供用水。公共供水设施包括正规的水厂和虽达不到水厂标准,但不是临时供水设施,水质符合标准的其他公共供水设施。

综合生产能力

指按供水设施取水、净化、送水、输水干管等环节设计能力计算的综合生产能力。计算时，以四个环节中最薄弱的环节为主确定能力。没有设计能力的按实际测定的能力计算。对于经过更新改造后，实际生产能力与设计能力相差很大的，按实际能力填报。

供水管道长度

指从送水泵至用户水表之间直径 75mm 以上所有供水管道的长度。不包括从水源地到水厂的管道、水源井之间的连络管、水厂内部的管道、用户建筑物内的管道和新安装尚未使用的管道。按单管计算，即：如在同一条街道埋设两条或两条以上管道时，应按每条管道的长度相加计算。

年供水总量

指一年内供给建成区范围内的全部水量，包括有效供水量和损失的水量。既包括镇（乡）自建的供水设施供应的水量，也包括市（县）或其他乡镇供应的水量。

其中：**年生活用水量**指居民家庭与公共服务的年用水量。包括饮食店、医院、商店、学校、机关、部队等单位生活用水量，以及生产单位装有专用水表计量的生活用量（不能分开者，可不计）。

年生产用水量指生产运营单位在生产、运营过程中的年用水量。

用水人口

指报告期末集中供水设施供给生活用水的家庭用户总人口数。可按本地居民户均人口数乘以家庭用水户数计算。

用气人口

指报告期末使用燃气（人工煤气、天然气、液化石油气）的家庭用户总人口数。可以本地居民平均每户人口数乘以燃气家庭用户数计算。

在统计液化气家庭用户数时，年平均用量低于 90 公斤的户数，忽略不计。

集中供热面积

指通过热网向建成区内各类房屋供热的房屋建筑面积。只统计供热面积达到 1 万平方米及以上的集中供热设施。

道路

指建成区范围内具有交通功能的各种道路。包括主干路、干路、支路、巷路。只统计路面宽度在 3.5 米（含 3.5 米）以上的各种铺装道路。

铺装道路路面包括水泥混凝土、沥青混凝土（柏油）、块石、砖块、混凝土预制块等形式。

道路长度、面积的计算

道路长度包括道路长度和与道路相通的桥梁、隧道的长度。道路面积只包括路面面积和与道路相通的广场、桥梁、停车场的面积，不含隔离带和绿化带面积。

污水处理厂

指污水通过排水管道集中于一个或几个处所，并利用由各种处理单元组成的污水处理系统进行净化处理，最终使处理后的污水和污泥达到规定要求后排放水体或再利用的生产场所。

污水处理厂主要有以下几种处理工艺：缺氧、好氧工艺（AO工艺）、2.厌氧、缺氧、好氧工艺（AAO工艺）、膜生物反应器工艺（MBR工艺）、氧化沟工艺（OD工艺）、序批式活性污泥工艺（SBR系列工艺）等。

氧化塘是污水处理的一种工艺，严格按《氧化塘设计规范》运行管理的氧化塘应作为污水处理厂。

污水处理装置

指在厂矿区设置的处理工业废水和周边地区生活污水的小型集中处理设备，以及居住区、度假村中设置的小型污水处理装置。

污水处理能力

指污水处理设施每昼夜处理污水量的设计能力，没有设计能力的按实际能力计算。

年污水处理总量

指一年内各种污水处理设施处理的污水量，包括将本地的污水收集到外地的污水处理设施处理的量，其中污水处理厂处理的量计为集中处理量。

绿化覆盖面积

指建成区内的乔木、灌木、草坪等所有植被的垂直投影面积。包括各种绿地的绿化种植覆盖面积、屋顶绿化覆盖面积以及零散树木的覆盖面积，不包括植物未覆盖的水域面积。

绿地面积

指报告期末建成区内用作园林和绿化的各种绿地面积。包括公园绿地、生产绿地、防护绿地、附属绿地的面积。其中：**公园绿地**指向公众开放的、以游憩为主要功能，有一定游憩设施的绿地。

生产绿地指为绿化提供苗木、花草、种子的苗圃、花圃、草圃等圃地。

防护绿地指用于安全、卫生、防风等功能的绿地。

附属绿地指建设用地中绿地之外各类用地中的附属绿化用地。包括居住用地、公共设施用地、工业用地、仓储用地、对外交通用地、道路广场用地、市政设施用地和特殊用地中的绿地。

统计时注意单位是"公顷"，1 公顷=1 万平方米。

年生活垃圾处理量

指一年将建成区内的生活垃圾运到生活垃圾处理场（厂）进行处理的量，包括无害化处理量和简易处理量。

公共厕所数量

指在建成区范围内供居民和流动人口使用的厕所座数。统计时只统计独立的公厕，不包括公共建筑内附设的厕所。

房屋

按使用性质可划分为住宅、公共建筑和生产性建筑。

其中：**住宅**指坐落在村镇范围内，上有顶、周围有墙，能防风避雨，供人居住的房屋。按照各地生活习惯，可供居住的帐篷、毡房、船屋等也包括在内，兼作生产用房的房屋可以算为住宅。包括厂矿、企业、医院、机关、学校的集体宿舍和家属宿舍，但不包括托儿所、病房、疗养院、旅馆等具有专门用途的房屋。

公共建筑指坐落在村镇范围内的各类机关办公、文化、教育、医疗卫生、商业等公共服务用房。

生产性建筑指坐落在村镇范围内的包括乡镇企业厂房、养殖厂、畜牧场等用房及各类仓库用房的建筑面积。

本年竣工建筑面积

指本年内完工的住宅（公共建筑、生产性建筑）建筑面积。包括新建、改建和扩建的建筑面积，未完工的在建房屋不要统计在内。统计时按镇（乡）建成区和村庄分开统计。

其中：**房地产开发**建筑面积，指房地产开发企业，按照城乡建设规划要求，立项审批（备案）并取得《施工许可证》后，在依法取得土地使用权的土地上开发的楼盘或小区工程项目的建筑面积。

年末实有建筑面积

指本年年末，坐落在村镇范围内的全部住宅（公共建筑、生产性建筑）建筑面积。统计时按镇（乡）建成区和村庄分开统计。计算方法为：

年末实有房屋建筑面积=上年末实有+本年竣工+本年区划调整增加-本年区划调整减少

-本年拆除（倒塌、烧毁等）

混合结构以上

指结构形式为混合结构及其以上（如钢筋混凝土结构、砖混结构）的房屋建筑面积。

Explanatory Notes on Main Indicators

Indicators in the statistics for Cities and County Seats

Population Density

It refers to the density of population in a given zone. The calculation equation is:

$$\text{Population Density} = \frac{\text{Population in Urban Areas} + \text{Urban Temporary Population}}{\text{Urban Area}}$$

Daily Domestic Water Use Per Capita

It refers to average amount of daily water consumed by each person. The calculation equation is

Daily Domestic Water Use Per Capita= (Water Consumption by Households + Water Use for Public Service+Domestic water consumption in the free water supply) ÷Population with Access to Water Supply ÷Calendar Days in Reported Period× 1000 liters

Water Coverage Rate

It refers to proportion of urban population supplied with water to urban population. The calculation equation is:

$$\text{Water Coverage Rate} = \frac{\text{Urban Population with Access to Water Supply (Including Temporary Population)}}{\text{Urban Permanent Population} + \text{Urban Temporary Population}} \times 100\%$$

$$\text{Public Water Coverage Rate} = \frac{\text{Urban Population with Access to Public Water Supply (Including Temporary Population)}}{\text{Urban Permanent Population} + \text{Urban Temporary Population}} \times 100\%$$

Gas Coverage Rate

It refers to the proportion of urban population supplied with gas to urban population. The calculation equation is:

$$\text{Gas Coverage Rate} = \frac{\text{Urban Population with Access to Gas (Including Temporary Population)}}{\text{Urban Permanent Population} + \text{Urban Temporary Population}} \times 100\%$$

Surface Area of Urban Roads Per Capita

It refers to the average surface area of urban roads owned by each urban resident at the end of reported period. The calculation equation is:

$$\text{Surface Area of Roads Per Capita} = \frac{\text{Surface Area of Roads in Given Urban Areas}}{\text{Urban Permanent Population} + \text{Urban Temporary Population}}$$

Density of Road Network in Built Districts

It refers to the extent which roads cover built districts at the end of reported period. The calculation equation is:

$$\text{Density of Road Network in Built Districts} = \frac{\text{Length of Roads in Built Districts}}{\text{Floor Area of Built Districts}}$$

Density of Drainage Pipelines in Built Districts

It refers to the extent which drainage pipelines cover built districts at the end of reported period. The calculation equation is:

$$\text{Density of Drainage Pipelines in Built Districts} = \frac{\text{Length of Drainage Pipelines in Built Districts}}{\text{Floor Area of Buil Districts}}$$

Wastewater Treatment Rate

It refers to the proportion of the quantity of wastewater treated to the total quantity of wastewater discharged at the end of reported period. The calculation equation is:

$$\text{Wastewater Treatment Rate} = \frac{\text{Quantity of Wastewater Treated}}{\text{Quantity of Wastewater Discharged}} \times 100\%$$

Centralized Treatment Rate of Wastewater Treatment Plants

It refers to the proportion of the quantity of wastewater treated in wastewater treatment plants to the total quantity of wastewater discharged at the end of reported period. The calculation equation is:

$$\text{Centralized Treatment Rate of Wastewater Treatment Plants} = \frac{\text{Quantity of wastewater treated in wastewater treatment facility}}{\text{Quantity of wastewater discharged}} \times 100\%$$

Public Recreational Green Space Per Capita

It refers to the average public recreational green space owned by each urban dweller in given areas. The calculation equation is:

$$\text{Public Recreational Green Space Per Capita} = \frac{\text{Public green space in given urban areas}}{\text{Urban Permanent Population} + \text{Urban Temporary Population}}$$

Green Coverage Rate of Built Districts

It refers to the ratio of green coverage area of built districts to surface area of built districts at the end of reported period. The calculation equation is:

$$\text{Green Coverage Rate of Built Districts} = \frac{\text{Green Coverage Area of Built Districts}}{\text{Area of Built Districts}} \times 100\%$$

Green Space Rate of Built Districts

It refers to the ratio of area of parks and green land of built districts to the area of built up districts at the end of reported period. The calculation equation is:

$$\text{Green Space Rate of Built Districts} = \frac{\text{Area of parks and green land of built districts}}{\text{Area of Built Districts}} \times 100\%$$

Domestic Garbage Treatment Rate

It refers to the ratio of quantity of domestic garbage treated to quantity of domestic garbage produced at the end of reported period. The calculation equation is:

$$\text{Domestic Garbage Treatment Rate} = \frac{\text{Quantity of Domestic Garbage Treated}}{\text{Quantity of Domestic Garbage Produced}} \times 100\%$$

Domestic Garbage Harmless Treatment Rate

It refers to the ratio of quantity of domestic garbage treated harmlessly to quantity of domestic garbage produced at the end of reported period. The calculation equation is:

$$\text{Domestic Garbage Harmless Treatment Rate} = \frac{\text{Quantity of Domestic Garbage Treated Harmlessly}}{\text{Quantity of Domestic Garbage Produced}} \times 100\%$$

Urban (county) District Area

It refers to the total land area (including water area) under jurisdiction of cities(counties).

Urban (county) Area

It refers to the area of a city's urban and county area.

Urban area of a city includes: (1) City-level areas administered by neighborhood office; (2) other towns (villages) connected to city public facilities, residential facilities and municipal utilities; (3) Independent Industrial and Mining District, Development Zones, special areas like research institutes , universities and colleges with permanent residents of 3000 above.

County area of a city includes: (1)Town or areas administered by neighborhood office; (2) other towns (villages) connected to county public facilities, residential facilities and municipal utilities; (3) Independent Industrial and Mining District, Development Zones, special areas like research institutes , universities and colleges with permanent residents of 3000 above in county.

Towns (villages) connected to city public facilities, residential facilities and municipal utilities means that towns and urban centers are connected with public facilities, residential facilities and municipal utilities that are constructed or under construction, and not cut off by non-construction land like water area, agricultural land, parks, woodland or pasture.

As to conurbation or cities organized in scattered form, urban area should include the separated or decentralized area.

Urban (county) Permanent Population

It refers to permanent residents in urban (county) areas, which are in compliance with the number of permanent residents registered in public security authorities.

Temporary Population

Temporary population includes people who leave permanent address, and live for over half a year in a place. Counted by urban area, county, urban area and county seat,which are in compliance with the number of temporary residents registered in public security authorities.

Area of Built District

It refers to large scale developed quarters within city (county) jurisdiction with basic public facilities and utilities. For a nucleus city, the built-up district consists of large-scale developed quarters with basic public facilities and utilities, which are either centralized or decentralized. For a city with several towns, the built-up district consists of several developed quarters attached in succession with basic public facilities and utilities. Range of built-up district is the area encircled by the limits of the built-up district, i.e. the range within which the actual developed land of this city exists.

Area of Urban Land for development Purpose

It refers to area of land including land for Residential Development, Administration and Public Services, Commercial and Business Facilities, Industrial, Manufacturing, Logistics and Warehouse, Road, Street and Transportation, Municipal Utilities, Green Space and Square.Separately counted by the planned construction land and the current construction land.

The area of land used for planned construction refers to the area of land used for cities (towns) as determined by the latest version of the overall urban plans approved according to law for the towns where the relevant cities and county people's governments are located at the end of the reporting period.

The area of current land for construction purposes refers to the area corresponding to the actual situation of urban land for construction purposes in the towns where the relevant cities and county people's governments are located at the end of the reporting period.

Urban Maintenance and Construction Fund

It refers to fund used for urban construction and maintenance, including urban maintenance and construction tax, extra-charges for public utilities, financial allocation from the central government and local governments, domestic loan, securities revenue, foreign investment, revenue from land transfer and assets replacement, self-raised fund by municipal utilities, charges by administrative and institutional units for urban maintenance and construction, pooled revenue and other revenues.

Urban Maintenance and Construction Tax

It is one of local taxes imposed according to *Temporary Regulations of People's Republic of China on Urban Maintenance and Construction Tax*. It is levied based on actual value-added tax, consumption tax and business tax paid by a taxpayer. It is also paid with value-added tax, consumption tax and business tax at the same time. Different rates apply to different places. The rate comes to 7% in urban areas, 5% in counties and towns and 1% beyond these places.

Extra Charge from Urban Utilities

It is additional revenue obtained from additional fees imposed on provision of products and service of such urban utilities as power generation, water supply for the industry, operation of public bus and trolley bus, taped water supply, electricity for lightening, telephone operation, gas and ferry etc. for the purpose of urban construction and maintenance.

Investment in Fixed Assets

It is the economic activities featuring construction and purchase of fixed assets, i.e. it is an essential means for social reproduction of fixed assets. The process of reproducing fixed assets includes fixed assets renovation (part and full renovation), reconstruction, extension and new construction etc. According to the new industrial financial accounting system, cost of major repairs for part renovation of fixed assets is covered by direct cost. According to the current investment management and administrative regulations, any repair and maintenance works are not included in statistics as investment in fixed assets. Innovation projects on current municipal service facilities should be included in statistics.

Newly Added Fixed Assets

They refer to the newly increased value of fixed assets, including investment in projects completed and put into operation in the reported period, and investment in equipment, tools, vessels considered as fixed assets as well as relevant expenses should be included in. Other construction expenses that increase the value of fixed assets should be included in the newly added fixed assets along with the project delivered for use.

Newly Added Production Capacity (or Benefits)

It refers to newly added design capacity through investment in fixed assets. Newly added production capacity or benefits is calculated based on projects which can independently produce or bring benefits once projects are put into operation.

Integrated Production Capacity

It refers to a comprehensive capacity based on the designed capacity of components of the process, including water collection, purification, delivery and transmission through mains. In calculation, the capacity of weakest component is the principal determining capacity. For the updated and reformed, filling in the new design capability after the updated and reformed.

Length of Water Pipelines

It refers to the total length of all pipes from the pumping station to individual water meters. If two or more pipes line in parallel in a same street, the length of water pipelines is the length sum of each line.

Total Quantity of Water Supplied

It refers to the total quantity of water delivered by water suppliers during the reported period, including accounted water and unaccounted water.

Accounted water refers to the actual quantity of water delivered to and used by end users, including water sold and free.

Unaccounted Water

It refers to sum of water leakage due to malfunction of water meters, measurement of water loss and other loss of water. The measurement of loss includes the total water loss for residential users and the water loss with table error for non-residential users; Other loss of water refers to the loss of water caused by management factors such as unregistered users' water and users' refusal to check.

Quantity of Fresh Water Used

It refers to the quantity of water obtained from any water source for the first time, including tap water, groundwater, surface water. As for a city, it includes the quantity of fresh water used by urban water suppliers and customers in different industries and sectors.

Among which, **the amount of new industrial water taken** refers to the amount of fresh water that is actually drawn from various water sources and used for any purpose in order to ensure the normal progress of industrial production and the need for water in the production process, including the amount of indirect cooling water and process water, new boiler water volume and other new water volume.

Quantity of Recycled Water

It refers to the sum of water recycled, reclaimed and reused by customers.

Among which, **the amount of recycled industrial water** refers to the sum of the amount of recycled water used in the domestic and production water of industrial enterprises and the amount of water recovered and reused directly or after treatment.

Quantity of Water Saved

It refers to the water saved in the reported period through efficient water saving measures, e.g. improvement of production methods, technologies, equipment, water use behavior or replacement of defective and inefficient devices, or strengthening of management etc.

Integrated Gas Production Capacity

It refers to the combined capacity of components of the process such as gas production, purification and delivery with an exception of the capacity of backup facilities at the end of reported period. It is usually calculated based on the design capacity. Where the actual capacity surpluses the design capacity, integrated capacity should be calculated based on the actual capacity, mainly depending on the capacity of the weakest component.

Length of Gas Supply Pipelines

It refers to the length of pipes operated in the distance from a compressor's outlet or a gas station exit to pipes connected to individual households at the end of reported period, excluding pipes through coal gas production plant, delivery station, LPG storage station, bottled station, storage and distribution station, air mixture station, supply station and user's building.

Quantity of Gas Supplied

It refers to amount of gas supplied to end users by gas suppliers at the end of reported period. It includes sales amount and loss amount.

Gas Stations for Gas-Fueled Motor Vehicles

They are designated stations that provide fuels such as compressed natural gas, LPG to gas-fueled motor vehicles. Statistics should be based on different gas types.

Heating Capacity

It refers to the designed capacity of heat delivery by heat suppliers to urban customers.

Total Quantity of Heat Supplied

It refers to the total quantity of heat obtained from steam and hot water, which is delivered to urban users by heat suppliers during the reported period.

Length of Heating Pipelines

It refers to the total length of pipes for delivery of steam and hot water from heat sources to entries of buildings, excluding lines within heat sources.

Among which, **the primary pipe network** refers to the heating pipeline from the heat source to the heating station, **the secondary pipe network** refers to the heating pipeline from the heating station to the user.

Urban roads

They refer to roads, bridges, tunnels and auxiliary facilities that are provided to vehicles and passengers for transportation. Urban roads consist of drive lanes and sidewalks. Only paved roads with width with and above 3.5 m, including roads within on-limits industrial parks and residential communities are included in statistics.

Length of Roads

It refers to the length of roads and bridges and tunnels connected to roads. It is calculated based on the length of centerlines of traffic lanes.

Surface Area of Roads

It refers to surface area of roads and squares, bridges and tunnels connected to roads (surface area of roadway and sidewalks are calculated separately). Surface area of sidewalks is the area sum of each side of sidewalks including pedestrian streets and squares, excluding car-and-pedestrian mixed roads.

Bridges

They refer to constructed works that span natural or man-made barriers. They include bridges spanning over rivers, flyovers, overpasses and underpasses.

Number of Road Lamps

It refers to the sum of road lamps only for illumination, excluding road lamps for landscape lightening.

Length of Flood Control Dikes

It refers to actual length of constructed dikes, which sums up the length of dike on each bank of river. Where each bank has more than one dike, only the longest dike is calculated.

Quantity of Wastewater Discharged

It refers to the total quantity of domestic sewage and industrial wastewater discharged, including effluents of sewers, ditches, and canals.

(1) It is calculated by multiplying the daily average of measured discharges from sewers, ditches and canals with calendar days during the reported period.

(2) If there is a device to read actual quantity of discharge, it is calculated based on reading.

(3) If without such ad device, it is calculated by multiplying total quantity of water supply with wastewater drainage coefficient.

Category of urban wastewater	Wastewater drainage coefficient
Urban wastewater	0.7-0.8
Urban domestic wastewater	0.8-0.9
Urban industrial wastewater	0.7-0.9

Length of Drainage Pipelines

The total length of drainage pipelines is the length of mains, trunks, branches plus distance between inlet and outlet in manholes and junction wells. The calculation should be based on a single pipe, that is, if there are two or more drainage pipes side by side on the same street, the calculation should be based on the length of each drainage pipe.

Among which, **Sewage Pipe** refers to a drainage pipe dedicated to discharge sewage.

Rainwater Pipe refers to drainage pipes dedicated to draining rainwater.

Rain and Sewage Confluence Pipeline refers to the drainage pipeline where rainwater and sewage enter the same pipeline for drainage at the same time.

Quantity of Wastewater Treated

It refers to the actual quantity of wastewater treated by wastewater treatment plants and other treatment facilities in a physical, or chemical or biological way.

Among which, **the treatment outside the city (county)** refers to the sewage treatment plant as a regional facility that not only treats the sewage of the city (county), but also treats the sewage of other cities, counties, or other villages outside the city (county). This part of the sewage treatment volume is separately counted and deducted when calculating the sewage treatment rate of the city (county).

Dry Sludge Production

It refers to the final quantity of dry sludge produced in the process of wastewater treatment plants. Dry sludge refers to the amount of sludge calculated by the mass of dry solids, with a moisture content of 0. If the moisture content of the produced wet sludge is n%, then the amount of dry sludge produced = the amount of wet sludge produced $\times$ (1-n%).

Dry Sludge Disposal Volume

It refers to the amount of dry sludge to be treated and disposed according to the standard during the reporting period. Fill in the statistics according to land use, building materials use, incineration, landfill and others.

Among which, **Sludge Land Utilization** refers to the disposal of sludge products after the treatment of the standard for landscaping, land improvement, woodland, agriculture and other occasions.

The Utilization of Sludge Building Materials refers to the disposal of the products after sludge treatment reaches the standard as part of the raw materials of building materials such as bricks and cement clinker.

Sludge Incineration refers to the treatment and disposal methods that use incinerators to completely mineralize sludge into a small amount of ashes, including individual incineration, and co-incineration with industrial kilns such as domestic waste and thermal power plants.

Land Filling of Sludge refers to a disposal method in which engineering measures are taken to pile, fill, and bury the sludge products that have reached the treatment standards, and place them in a controlled site.

Green Coverage Area

It refers to the vertical shadow area of vegetation such as trees (arbors), shrubs and grasslands. It is the the vertical shadow area sum of green space, roof greening and scattered trees coverage. The vertical shadow area of bushes and herbs under the shadow of trees and herbs under the shadow of shrubs should not be counted repetitively.

Area of Green Space

It refers to the area of all spaces for parks and greening at the end of reported period, which includes area of public recreational green space, shelter belt, land for squares, attached green spaces, and the area of regional green space in built district.

Among which, **Public Recreational Green Space** refers to green space with recreation and service facilities. It also serves some comprehensive functions such as improving ecology and landscape and preventing and mitigating disasters.

Shelter Belt refers to the green space that is independent of land and has the functions of hygiene, isolation, safety, and ecological protection, and is not suitable for tourists to enter. It mainly includes protective green space for health isolation, protective green space for roads and railways, protective green space for high-pressure corridors and protective green space for public facilities.

Land for Squares refer to the urban public activity space with the functions of recreation, memorial, gathering and avoiding danger.

Attached Green Spaces refer to green land attached to various types of urban construction land (except "green land and square land"). Including residential land, land for public management and public service facilities, land for commercial service facilities, industrial land, land for logistics and storage, land for roads and transportation facilities, land for public facilities and other green spaces.

Regional Green Space refers to the green land located outside the urban construction land, which has the functions of urban and rural ecological environment, natural resources and cultural resources protection, recreation and fitness, safety protection and isolation, species protection, garden seedling production and so on.

Parks

They refer to places open to the public for the purposes of tourism, appreciation, relaxation, and undertaking scientific, cultural and recreational activities. They are fully equipped and beautifully landscaped green spaces. There are different kinds of parks, including general park, Children Park, park featuring historic sites and culture relic, memorial park, scenic park, zoo, botanic garden, and belt park. Recreational space within communities is not included in statistics, while only general parks, theme parks and belt parks at city or district level are included in statistics.

Among which, **Free-ticket Parks** refer to parks that are free to the public and do not sell tickets.

Area of Roads Cleaned and Maintained

It refers to the area of urban roads and public places cleaned and maintained at the end of reported period, including drive lanes, sidewalks, drive tunnels, underpasses, green space attached to roads, metro stations, elevated roads, flyovers, overpasses, squares, parking areas and other facilities. Where a place is cleaned and maintained several times a day, only the time with maximum area cleaned is considered.

Among which, **Mechanized Road Cleaning and Cleaning Area** refers to the road area that was cleaned and cleaned with large and small machinery such as road sweepers (machines) and washing vehicles at the end of the reporting period. When multiple machines are repeatedly used on a road, only the area cleaned and cleaned by one machine is calculated, and the statistics cannot be repeated.

Quantity of Domestic Garbage and Construction Waste Transferred

It refers to the total quantity of domestic garbage and construction waste collected and transferred to treatment grounds. Only quantity collected and transferred from domestic garbage and construction waste sources or from domestic garbage transfer stations to treatment grounds is calculated with exception of quantity of domestic garbage and construction waste transferred second time.

Food waste is a part of domestic waste. Whether it is removed separately or mixedly, it should be counted in the volume of domestic waste.

Among which, **the amount of food waste removal and disposal** refers to the total amount of food waste that is separately removed and disposed of separately, excluding the part that is mixed in the domestic waste.

Latrines

They are used by urban residents and flowing population, including latrines placed at both sides of a road and public place. They usually consist of detached latrine movable latrine and attached latrine. Only detached latrines rather than latrines attached to public buildings are included in the statistics.

Detached latrine is classified into three types. Moveable latrine is classified into fabricated latrine, separated latrine, motor latrine, trail latrine and barrier-free latrine.

Specific Vehicles and Equipment for Urban Environment Sanitation

They refer to vehicles and equipment specifically for environment sanitation operation and supervision, including vehicles and equipment used to clean, flush and water the roads, remove snow, clean and transfer garbage and soil, environment sanitation monitoring, as well as other vehicles and equipment supplemented, such as garbage truck, road clearing truck, road washing truck, sprinkling car, vacuum soil absorbing car, snow remover, loading machine, bulldozer, compactor, garbage crusher, garbage screening machine, salt powder sprinkling machine, sludge absorbing car and special shipping. Vehicles and equipment for long-term lease are also included in the statistics.

The number of special vehicles for road cleaning and cleaning and the special vehicles for domestic garbage transportation are separately counted.

Indicators in the Statistics for Villages and Small Towns

Population density

Population density is the measure of the population per unit area in Built districts.

Calculation equation is:

Population Density in Built Districts(People per square kilometer)= Permanent Population in Built Districts(People) ×100/ Area of Built Districts(Hectares)

Water Coverage Rate

It refers to proportion of population supplied with water in built districts (villages) to population in built districts (villages) at the end of reported period. Statistics is made by built district and village respectively. The calculation equation is:

Water Coverage Rate in Built Districts(%) = Population with Access to Water Supply in Built Districts(People)/ Permanent Population in Built Districts(People)×100%

Water Coverage Rate in Villages(%) = Population with Access to Water Supply in Villages(People)/ Permanent Population in Village(People)×100%

Gas Coverage Rate

It refers to the proportion of population supplied with gas in built districts to urban population in built districts at the end of reported period. The calculation equation is:

Gas Coverage Rate(%)=Population Supplied with Gas in Built Districts(People)/Permanent Population in Built Districts(People)×100%

Surface Area of Roads Per Capita

It refers to the average surface area of roads owned by each urban resident in built district at the end of reported period. The calculation equation is:

Surface Area of Roads in Built Districts Per Capita(Square meters/person)= Surface Area of Roads in Built Districts(Ten thousand square meters)×10^4 / Permanent Population in Built Districts(people)

Density of Drainage Pipeline Culverts

It refers to the extent which drainpipes cover built districts at the end of reported period. The calculation equation is:

Density of Drainage Pipeline Culverts in Built Districts(Kilometers/square kilometers)=Length of Drainage Pipelines in Built Districts(Kilometers)+Length of Drainage Culverts in Built Districts(Kilometers)×100/Floor Area of Built Districts(Hectares)

Public Recreational Green Space Per Capita

It refers to the average public recreational green space owned by each dwell in built district at the end of reported period. The calculation equation is:

Public Recreational Green Space in Built Districts Per Capita (Square meters/person)= Area of Public Recreational Green Space in Built Districts(Hectares)×10^4/ Permanent Population in Built Districts(people)

Green Coverage Rate in Built Districts

It refers to the ratio of green coverage area of built districts to surface area of built district at the end of reported period. The calculation equation is:

Green Coverage Rate in Built Districts(%) = Built Districts Green Coverage Area(Hectares) / Area of Built Districts(Hectares) ×100%

Green Space Rate in Built Districts

It refers to the ratio of area of parks and green land of built districts to the area of built districts at the end of reported period. The calculation equation is:

Green Space Rate in Built Districts(%)=Built Districts Area of Parks and Green Space(Hectares) /Area of Built Districts(Hectares)×100%

Daily Domestic Water Use Per Capita

It refers to average amount of daily water consumed by each person. The calculation equation is:

Domestic Water Use Per Capita= Domestic Water Consumption / Population with Access to Water Supply / Calendar Days in Reported Period× 1000 liters

Built District

It refers to large scale developed quarters within city jurisdiction with basic public facilities and utilities, The scope of the built-up area generally refers to the area covered by the outline of the built-up area, that is, the range reached by the actual construction land which is subject to area verified by local construction authorities or planning authorities.

Village

It refers to a collection of houses and other buildings in a country area where rural residents live and produce.

Because there is a big gap between the town (township) government station and the municipal public facilities in the village, we divide the special areas such as town, township and farm into built-up areas and villages for statistics in the report form.

Villages Planning

It addresses, under guidance of town overall planning or township planning, detailed arrangement for residential buildings, water supply, power supply, roads, greening, environmental sanitation and supplementary facilities for production according to local economic development level. Development Planning in Villages must be approved by the superior agreed to implement. Special attention should be given to planning validation in statistics.

Input in Village and Town Planning Development this Year

It refers to the total investment in the preparation of various plans such as overall planning, special planning, detailed planning, and village planning in the town (township) this year.

Public Water Supply

It refers to water supply for production, domestic use as well as public service by public water suppliers through water supply pipes and accessories. Public water supply facilities include regular plants , and other facilities which are not temporary facilities and produce water in compliance with standards.

Integrated Water Production Capacity

It refers to a comprehensive capacity based on the design capacity of components of the process, including water collection, purification, delivery and transmission through mains. In calculation, the capacity of weakest component is the principal determining capacity. Without design capacity, it is calculated based on capacity measured. If real production capacity differs widely from design capacity through renovation, real production capacity will be the basis for calculation.

Length of Water Pipelines

It refers to the total length of all pipes from the pumping station to individual water meters, excluding pipes from water resources to water plants, connection pipes between water resources wells, pipes within water plants and end-use buildings as well as pipes newly installed but not uses yet. If two or more pipes line in parallel in a same street, the length of water pipelines is the length sum of each line.

Total Annual Water Supply

It refers to the total amount of water supplied to the built-up area within a year, including the effective water supply and the amount of water lost. It includes not only the amount of water supplied by the self-built water supply facilities of the town (township), but also the amount of water supplied by the city (county) or other towns.

Among which, **Yearly Domestic Water Use** refers to amount of water consumed by households and public facilities, including domestic water use by restaurant, hospital, department store, school, government agency and army as well as water metered by special meters installed in industrial users for domestic use.

Yearly Water Consumption for Production refers to water consumed each year by industrial users for production and operation.

Population with Access to Water

It refers to the total number of household users whose water was supplied by centralized water supply facilities at the end of the reporting period. It can be calculated by multiplying the number of household water users by the average number of family members per household.

Population with Access to Gas

It refers to total quantity of customers in households supplied with gas (including man-made gas, natural gas and LPG) at the end of reported period.

When comes to number of households supplied with LPG, the households with yearly average use of LPG being below 90 kilograms are not included in statistics.

Area of Centrally Heated District

It refers to floor area of building stock in built-up districts supplied with centralized heating system. Only areas with centrally heated floor space of 10,000 square meters and above are included in statistics.

Road

It refer to a variety of roads with traffic functions within the built-up area. Including trunk road, trunk road, branch road, lane road. Roads whose width is 3.5 meters or over are included in statistics.

Pavement road surface includes cement concrete, asphalt concrete (asphalt), block stone, brick, concrete precast blocks and other forms.

Length and Surface Area of Road

It refers to the length of roads and bridges and tunnels connected to roads. It is calculated based on the length of centerlines of traffic lanes. Surface area of roads only includes surface area of roads and squares, bridges and tunnels connected to roads, excluding separation belt and green belt.

Wastewater Treatment Plant

It refers to places where wastewater is collected in one or a few places through drainage pipes, then purified and treated by wastewater treatment system with treated water and sludge being discharged or recycled in accordance with related standards and requirements.

Sewage treatment plants mainly have the following treatment processes: anoxic, aerobic process (AO process), 2. Anaerobic, anoxic, aerobic process (AAO process), membrane bioreactor process (MBR process), oxidation ditch process (OD process), sequence batch activated sludge process (SBR series process), etc.

Oxidation pond is one of techniques applied in wastewater treatment. Oxidation pond which is strictly operated in accordance with Oxidation Pond Design Standard should be employed as wastewater treatment plant

Wastewater Treatment Facilities

They refer to small-size and centralized equipment and devices installed in plant or mining area to treat industrial wastewater and domestic wastewater in surrounding areas , as well as wastewater equipment and devices placed in residential areas.

Wastewater Treatment Capacity

It refers to designed capacity for amount of wastewater treated every day and night by wastewater treatment facilities. If there is lack of design capacity, real capacity will be the determining factor for calculation.

Yearly Amount of Wastewater Treated

It refers to amount of wastewater treated by different wastewater treatment facilities within a year, including the amount of local wastewater collected and delivered to wastewater treatment facilities in other areas. The amount of wastewater treated in wastewater treatment plants are regarded as centralized treated amount.

Green Coverage Area

It refers to the vertical shadow area of vegetation such as trees (arbors), shrubs and grasslands. It is the area sum of green space, roof greening and scattered trees coverage. The surface area of water bodies which are not covered by green is not included.

Area of Green Space

It refers to the area of all spaces for parks and greening at the end of reported period, which includes area of public recreational green space, green land for production, shelter belt, and attached green spaces.

Among which, **Public Recreational Green Space** refers green space equipped with recreational facilities and open to the public for recreational purposes.

Green Land for Production refers to the nursery, flower and grass beds that provide seedlings, flowers and seeds for greening.

Shelter Belt refers to the green space used for safety, health, wind protection and other functions.

Attached Green Spaces refer to the ancillary green land in all kinds of land other than green land in the construction land. Including residential land, land for public facilities, land for industry, land for storage, land for external traffic, land for roads and squares, land for municipal facilities and green land in special land.

Note that the unit of statistics is "hectare",1 hectare = 10,000 square meters.

Yearly Amount of Domestic Garbage Treated

It refers to amount of domestic garbage in built district which are delivered to domestic garbage treatment plants for harmless treatment or simplified treatment each year.

Number of Latrines

It refers to number of latrines for residents and floating population. Statistics only includes detached latrines, and latrines located within buildings are not included.

House

According to the nature of use, it can be divided into residential building, public building and building for production.

Among which, **Residential Building** refers to residential building with roof and wall that shelters people from external environment. In light of the local customs, houses also include tent, yurt and ship where people could live, building used for production in addition to habitation, and living quarters for workers and staff supplied by factories and mines, enterprises, hospitals, government agencies and schools. Nursery, ward, sanatorium, and hotel which are special purpose buildings are not included.

Public Building refers to buildings for government agencies, culture activities, healthcare, and business and commerce, which are located in towns and villages.

Building for Production refers to buildings for township enterprises, farms, stock farms, and warehouses located in towns and villages.

Floor Area of Buildings Completed this Year

It refers to floor area of buildings (public buildings and buildings for production) completed in this year, including floor area of new construction, renovation, and extension. On going construction is not included in statistics. Data are separately collected for the built-up districts of small towns and villages.

Among which, **Real Estate Development Building Area** refers to the construction of real estate or community engineering projects developed on land where land use rights are legally obtained by real estate development enterprises, in accordance with the requirements of urban and rural construction planning, after project approval (recording) and obtaining the Construction Permit area.

Real Floor Area of Buildings at the End of Year

It refers to floor area of total buildings (public buildings and buildings for production) located in towns and villages at the end of year. Calculation equation:

Real Floor Area of Buildings at the End of Year = Real Floor Area of Buildings at The End of Last Year+ Added Floor Area of Buildings Due to District Readjustment This Year- Reduced Floor Area of Buildings Due to District Readjustment This Year-floor Area of Buildings Dismantled (collapsed or burned down)This Year.

Building with Mixed Structure

It refers to floor area of buildings with mixed structure, such as reinforced concrete structure, brick-and-concrete structure etc.